Being and Value in Technology

Enrico Terrone • Vera Tripodi
Editors

Being and Value in Technology

palgrave
macmillan

Editors
Enrico Terrone
Department of Antiquity, Philosophy
and History
University of Genoa
Genoa, Italy

Vera Tripodi
Department of Philosophy
University of Milan
Milan, Italy

ISBN 978-3-030-88792-6 ISBN 978-3-030-88793-3 (eBook)
https://doi.org/10.1007/978-3-030-88793-3

This Palgrave Macmillan imprint is published by the registered company Springer Nature Switzerland AG.
The registered company address is: Gewerbestrasse 11, 6330 Cham, Switzerland

Acknowledgments

We would like to thank the authors of the papers collected in this volume, for their willingness to cooperate during the whole review process, as well as the referees, for carefully reading the contributions and for giving constructive comments which helped improving the volume. We are also grateful to the editorial committee of *Palgrave Macmillan* for all the support given to the publication of this book.

Contents

Notes on Contributors

Christopher Bartel is Professor of Philosophy at Appalachian State University and an Adjunct Research Fellow at Charles Sturt University. His research interests primarily lie within ethics and aesthetics, with a special focus on music and videogames. He is the author of *Video Games, Violence, and the Ethics of Fantasy: Killing Time* (Bloomsbury 2020).

Dana S. Belu is Professor of Philosophy in the Philosophy Department and Women's Studies Department at California State University, Dominguez Hills. She works at the intersection of phenomenology, feminist philosophy of technology and bioethics, and is the author of *Heidegger, Reproductive Technology & The Motherless Age* (Palgrave 2017).

Marco Fasoli is a Research Fellow at the Scuola Universitaria Superiore IUSS of Pavia. His research areas are the philosophy of cognitive sciences and the philosophy of technology. In recent years, he has focused on the ontological and moral issues surrounding cognitive artifacts, also addressing the impact of new digital technologies on personal well-being.

Manuel García-Carpintero obtained his PhD at the University of Barcelona and has taught there since then. He works in the philosophy of language and mind, and related epistemological and metaphysical issues. He is completing a book under contract with OUP on the nature of speech acts in general and assertion in particular, entitled *Tell Me What You Know*.

Wybo Houkes is Professor of Philosophy of Science and Technology at the Department of Industrial Engineering & Innovation Sciences of the Eindhoven University of Technology (TU/e). After studying theoretical physics and philosophy in Amsterdam and Leiden, he defended a PhD thesis on Kantianism in early twentieth-century philosophy at the University of Leiden (2000). Until 2004, he was a Post-doctoral Researcher in the program "The Dual Nature of Technical Artefacts" funded by the Netherlands Organization for Scientific Research (NWO). In 2008, he was awarded a Vidi fellowship for a project on evolutionary theories of technology. His research interests include technical artefacts and their functions; cultural-evolutionary models of technology; scientific modelling; and the interrelations between technology, individual users and societal practices. He (co-)authored three books, including Houkes/Vermaas, *Technical Functions* (Dordrecht: Springer, 2010), and articles in peer-reviewed journals such as *The British Journal for the Philosophy of Science*, *Design Studies*, *The Monist*, *Philosophy of Science*, *Research Policy*, *Studies in History and Philosophy of Science* and *Synthese*.

Enrico Terrone is Associate Professor at University of Genoa, Italy. He works on philosophical issues concerning aesthetics, ontology and technology. His main research is philosophy of film. His last book (co-authored with Luca Bandirali) is *Concept TV. An Aesthetics of Television Series*.

Vera Tripodi is an Assistant Professor at University of Milan, Italy. She specializes in feminist philosophy and ethics, bioethics, ethics of technology and social ontology. She is a Founding Member and Vice President of the SWIP ITALIA (The Society for Women in Philosophy—Italy).

Nele Van de Mosselaer is a Postdoctoral Researcher in Philosophy at the University of Antwerp (Belgium). Her research focuses on topics in philosophy of fiction, game studies, philosophy of technology and narratology. She has published on the emotional experience of fiction, the aesthetics of videogames and interactions with virtual entities. In her ongoing research, she investigates how the boundaries between reality and fiction are blurred within experiences of virtually represented environments.

Janna Van Grunsven is an Assistant Professor in TU Delft's ethics and philosophy of technology section and the university-wide coordinator of TU Delft's ethics and philosophy education. She is also a Research Fellow in the NWO-Gravitation research programme "The Ethics of Socially Disruptive Technologies." Her work, which often uses insights from 4E Cognition, has appeared in journals such as *Social Epistemology*, *Ethics and Information Technology*, *Techné: Research in Philosophy and Technology*, *Phenomenology and the Cognitive Sciences* and the *Journal of Consciousness Studies*.

Introduction

Technical artifacts are objects towards which one can take two different stances. First, one can take a descriptive stance, considering how a certain artifact is made, the way in which it functions, what it is for. Second, one can take an evaluative stance, wondering, for instance, whether a certain artifact is good or bad, or whether it is beautiful or ugly. From a philosophical perspective, the descriptive stance leads one to an ontological investigation concerning what technical artifacts are, while the evaluative stance leads one to consider whether technical artifacts are ethically problematic or rather possess moral qualities, as well as whether and how technical artifacts can be assessed aesthetically, finding them, for instance, beautiful or ugly.

This book proposes a philosophical approach to technology that aims to combine descriptive and evaluative stances. To wit, the descriptive inquiry into the being of technical artifacts in the first part of the volume paves the way for the evaluative inquiry into their value in the second and the third part. Understanding the nature of technical artifacts will put us in a privileged position for assessing the activities they enable and the experiences they elicit.

The first part thus concerns the ontology of technology. In *What Are Technical Artefacts in Patent Practice? A Practice-Based Ontology*, Wybo Houkes proposes a general framework for investigating the being of technical artifacts, while Enrico Terrone in *The Cyberspace Strikes Back. An Ontological Account of Social Networks* and Marco Fasoli in *Cognitive Artifacts between Cognitive Sciences and the Philosophy of Technology* focus on special kinds of technical artifacts, drawing connections between their ontological description and the possibility of their evaluation. Specifically, Terrone proposes an ontology of social networks which ends up individuating the special kind of experiences that those artifacts are meant to elicit, thereby paving the way for an evaluation of those artifacts on the basis of the evaluation of these experiences. Likewise, Fasoli's descriptive inquiry into the ontology of cognitive artifacts ends up in an evaluation of the ethical issues that those artifacts raise.

The second part is about the ethics of technology. The three essays of this part evaluate, from a moral perspective, contemporary technologies that are meant to reshape the basic biological features of mating and reproduction. In *Anticipating Sex Robots: A Critique of the Sociotechnical Vanguard Vision of Sex Robots as 'Good Companions',* Janna van Grunsven focuses on the moral significance of sex robots. In *The Right and Unfair Aspects of Artificial Womb Technology*, Vera Tripodi considers on the ethical valence of artificial wombs. In *Missed Opportunities: Feminist Grounds for Regulating Transnational Surrogacy, in the Anthropocene*, Dana Belu investigates the moral implications of assisted reproductive technology, focusing on commercial surrogacy.

The third part is about the aesthetics of technology. The three essays of this part concern the aesthetic appreciation of objects created by contemporary technologies that seem to challenge the traditional conception of art. In *Computer Art, Technology, and the Medium*, Christopher Bartel discusses the status of computer art, as well as the aesthetic responses that it is meant to elicit. In *Breaking the Fourth Wall in Videogames*, Nele van de Mosselaer argues that the peculiar aesthetic response that works of representational art such as plays and films elicit when they break the invisible barrier separating the world portrayed from the audience can be found also in videogames. In *Games, Artworks, and Hybrids*, Manuel García-Carpintero shows that videogames can elicit both ludic experiences, in virtue of their functioning as technically designed games, and aesthetic experiences, in virtue of their functioning as works of art.

Part I

Ontology

What Are Technical Artefacts in Patent Practice? A Practice-Based Ontology

Wybo Houkes

Constructing an ontology of technical artefacts is a curious endeavor. From a practical perspective, questions such as "What is my iPhone X?" or "Is there such a thing as my Tesla Model Y?" seem pointless—or a sign of deep alienation. The existence of smartphones appears as undeniable as Descartes' *cogito*, if not more so. Yet much work on the ontology of artefacts has consisted of doubts concerning their existence and attempts to dispel those doubts. Perhaps most famously, Peter Van Inwagen (1990) has argued that an artefact is an arrangement of 'simples': tables, tablets, and Teslas are elementary particles, arranged table-, tablet-, or Tesla-wise.

This chapter presents one more attempt to avoid—or perhaps side-step—this conclusion. Before doing so, however, it needs to be clarified what might have brought ontologists like Van Inwagen to their denial of artefacts. This suggests that a more constructive account of the ontology of artefacts may require a different method or 'meta-ontology'. In Sect. 1, the currently dominant (Quinean) meta-ontology is presented, together

W. Houkes (✉)
Eindhoven University of Technology, Eindhoven, The Netherlands
e-mail: w.n.houkes@tue.nl

E. Terrone, V. Tripodi (eds.), *Being and Value in Technology*,
https://doi.org/10.1007/978-3-030-88793-3_1

with the arguments that it generates for doubting the existence of artefacts. Then, we turn to an alternative, 'easy' approach, and reasons are given for exploring a variant that grounds ontologies in societal practices.

After this meta-ontological part, an ontology of technical artefacts is developed on the basis of a societally vital practice that governs our involvement with technologies, namely patent law. Section 2 sketches some key elements of patent law, focusing on the European Patent Convention (EPC). These elements create a context in which technical artefacts need to be distinguished from other types of entities, and from each other. Moreover, non-trivial conditions are imposed—both in the letter of the law and in jurisprudence—on what counts as an artefact. In Sect. 3, it is shown how these aspects of legal practice can be reconstructed by taking technical artefacts to be abstract types, with some distinctive, non-standard characteristics. Section 4 concludes.

1 Toward a Practice-Based Meta-ontology

Before developing an ontology of technical artefacts (henceforth, where possible: artefacts), it needs to be clear why such a project is called for, and what shape it should take. This requires, first of all, explaining why some have denied that there are artefacts: this is the background against which the more constructive project will be undertaken. Secondly, it is explained how philosophers in the analytic tradition deal with ontological issues. This section is therefore an excursion into meta-ontology, that is, the study of the aim and method of ontology, which is the theory of being. I will first sketch the currently dominant, Quinean meta-ontology and review how it can be employed to generate or substantiate doubts concerning the existence of artefacts (Sect. 1.1), before turning to an alternative, 'easy' approach for which I propose a practice-based variant (Sect. 1.2).

1.1 The Troubled (Non-)Existence of Artefacts

Ontology in the analytic tradition is rooted in an approach that originated with W.V.O. Quine's essay "On What There Is" (1948). This Quinean[1] meta-ontology is so dominant that it can be presented as 'The Standard View' (Berto & Plebani, 2015, especially Chaps. 2 and 3).

For Quine, the aim of ontology is to determine what there is, that is, to make a complete inventory of entities. Whereas such an inventorying task usually occurs in a clear practical or theoretical context (e.g., cataloguing a museum or identifying virus strains), the Quinean approach offers a formal criterion for 'ontological commitment' instead: to determine what there is, one should inspect language use, in particular the domains of quantification of statements. Here, paraphrasal into some formal language—for Quine, first-order predicate logic—is needed to reveal how statements may be true without requiring belief in all manner of objects. In addition, Quineans insist on the importance of *identity criteria*—through the slogan "No entity without identity". Types of entities should have clear (ideally, formalizable) conditions under which they are identical and non-identical. Often, these criteria are distinguished from *individuation* conditions, that is, those by which we identify entities as the same or different in actual practice.

In its application, the Quinean approach is not so much an ordinary-language-oriented inventorying enterprise as a massively reconstructive sanitization program. In Quine's words, language use resembles a "slum ... [that] is a breeding ground for disorderly elements" (1948, p. 4). Insistence on identity criteria should, aided by paraphrasal, lead to dismissal of many objects to which everyday statements appear to refer. Prominent examples of 'disorderly elements' that have been 'sanitized' by some Quineans include numbers, mental states, propositions, organisms, and technical artefacts. There is, it should be stressed, no consensus among Quineans about the ontological commitments on which we may forego. In every case, if there are arguments for doubting the existence of some objects, the 'sanitization' method is to reveal by paraphrasing how

[1] Sometimes called 'neo-Quinean' to indicate differences between the original approach and more recent successors. Since I will largely ignore those here, I use the simpler label.

ordinary language does not, in fact, require commitment to these objects (see, e.g., Varzi, 2007).

A comprehensive overview of the arguments that have led some Quineans to doubts regarding artefacts cannot be given here. Instead, I will briefly review the kinds of concerns that have been raised, also as an introduction to the types of arguments that characterize analytic ontology.

An artefact could be a material object, like a hedgehog; a mental one, like my intention to finish this paper before the deadline; or an abstract one, like the number five. For purposes of this review, assume that each option is plausible, that they are mutually exclusive and that, together, they are exhaustive. Quinean doubts, focusing in particular on identity criteria, can then be raised for each option—leaving artefacts without an ontological safe haven.

First, suppose that an artefact is a material object. Then, problems arise from the mere fact that engineering is not *creatio ex nihilo*: virtually every artefact is, on our best current understanding, composed of other (material) objects. Now, according to many—though definitely not all—Quineans, the identity criteria for material objects refer to their spatiotemporal location. Thus, the locations objects share must be identical: there is no reason to admit them both into our inventory. Because artefacts are composed of other objects, they would ultimately coincide with a collection of 'simples': material objects that have no constitutive parts. For practical purposes, we may refer to such collections as 'smartphones' or 'Teslas', but strictly speaking, talk of such objects is a convenient fiction and can be paraphrased away (Van Inwagen, 1990; Rosen & Dorr, 2002; Renz, 2016).

Some authors resist this conclusion and allow composite objects into their ontology, but only if we refer to them by *sortal predicates*—roughly, terms that categorize entities into sorts or kinds. At face value, this may seem overly permissive: after all, not only a term such as 'smartphone' serves this purpose, but so does 'unicorn'. To avoid this, so-called 'neo-Aristotelean' ontologies focus on those sortal predicates that, for a kind of entities, specify the essence or provide persistence conditions, that is, criteria for determining when something of the kind comes into being or ceases to exist. Leaving aside details and difficulties of the resulting accounts (e.g., Lowe, 1998; Wiggins, 2001), it should be noted that these ontologies are hardly more welcoming to technical artefacts than some neo-Quinean ones. Artefact kind terms are not considered genuine sortals, because they do not provide sufficiently robust individuation and persistence conditions. To quote these worries, expressed by two prominent neo-Aristoteleans instead of their arguments: “Can a *table* ‘morph’ into a *chair*? It seems so—in fact, it can do so very easily” (Lowe, 2014, p. 22) and “A clock is any time-keeping device, a pen is any rigid ink-applying writing implement, and so on” (Wiggins, 2001, p. 87). The deeper worry here is different from the composition problem mentioned above, but also refers to a basic fact of human dealings with technology: we can arguably enlist all manners of entities for practical purposes, and are only minimally constrained by any intrinsic features or ‘characteristic activities’ of these entities. Consequently, the items in especially broad functional categories or ‘instrumental kinds’ (Houkes & Vermaas, 2014) such as pens and tables, are so variegated that it becomes difficult to see what makes anything into a pen apart from our using it as such.

This may be reason to consider the second option: an ontology of artefacts as *mental* objects, for instance the content of mental states of such beliefs and intentions. After all, the previous considerations suggest that we tell artefacts apart, assign them to the same category, and determine how they come and cease to be by referring first and foremost to human intentions—those of designers or of users. Still, the Quinean meta-ontology can be readily employed to raise some steep obstacles for this ‘mental’ route. For one thing, it is difficult to state the identity conditions of mental states and their content: it is unclear when my representation of a pen is the same as yours; and even mine is hardly stable over time. For another, mental objects seem insufficiently persistent. A designer might manage to keep a particular artefact in mind throughout work hours or, less plausibly, throughout a design project, but artefacts intuitively persist after their designers or any individual user has focused on other things or, to make the point in a more extreme way, passed away. Thirdly, our engagement with technical artefacts is often collaborative or collective. An ontology in which, say, a soccer ball is primarily a mental object in the mind of a player would have a hard time analyzing how they manage to play together. These considerations may not fatally discredit mental objects or artefacts conceived as mental objects, but they may explain why hardly anyone has developed an ontology of artefacts along these lines.

This leaves the third option: that artefacts are abstract objects. To make this plausible, it is worthwhile to compare technical artefacts to art works, entities with which they share a long conceptual history. Art works have been analyzed as abstract objects in a large body of literature (e.g., Currie, 1989; Levinson, 1980). The arguments typically employ what has been called the 'Way of Negation' (Lewis, 1986). For instance, they involve denying that works of art have a definite spatiotemporal location—for instance, because they can have multiple copies or instantiations (e.g., renditions of *Bohemian Rhapsody*), persist despite damage to any such instantiation (e.g., Venus de Milo), and arguably precede and succeed any instantiation. Since neither material nor mental objects have such persistence conditions, only abstract objects fit the bill.

As forceful as these arguments may be—and we shall return to them later—they need not satisfy Quineans. Traveling down the Way of Negation implicitly assumes that works of art—or technical artefacts, for that matter—have to be *some* kind of entity. Thereby, it ignores the option to dismiss them from our inventory entirely. Arguing that a kind of entity cannot be identified by its spatiotemporal location is not tantamount to offering clear alternative identity or persistence conditions—so Quineans are free to respond that the arguments only show that artefacts are not physical objects; they need not be convinced that they are real objects at all. The quest to specify these alternative conditions, and thus offer a *positive* argument, is still ongoing and seldom focused on technical artefacts rather than, for instance, works of art or works in general (see Sect. 3 below). Without attempts at spelling out identity criteria, however, artefacts are easy victims of Quinean sanitization efforts.

1.2 A Practice-Based, Easy Alternative

That the Quinean approach can be employed to generate or substantiate doubts about the existence of artefacts—whether they be material, mental, or abstract—need not be bad news for your smartphone. It could just as well be a reason to look for an alternative meta-ontology that is better suited to account for our statements about and dealings with artefacts. One alternative, the neo-Aristotelean approach, was mentioned in passing above and found not to be more welcoming to artefacts.[2] Recently, yet another meta-ontology has gained significant popularity: a 'deflationist' or 'easy' approach to ontology. After outlining this, focusing on Amie Thomasson's recent (2015) version and ignoring differences between easy approaches, a practice-based variant is proposed as a basis for the ontology developed in this paper.

The easy approach is rooted in Carnap's (1950) distinction between internal and external existence questions. The former may be asked within a 'linguistic framework', governed by specific rules, and answered *within* this framework. An example is "Is there a prime number greater than 100?", which can be answered within the framework of number theory. By contrast, the question "Do numbers exist?" is not internal to any such framework. It is therefore either an unanswerable pseudo-question for lack of an appropriate framework, or rather a practical, evaluative question why we ought to accept a framework (in this case: number theory) in the first place. On any reading, constructing ontologies requires no substantive philosophical argumentation: internal questions can be answered through empirical or logical means within the relevant linguistic frameworks, and external questions require linguistic decision-making, weighing pros and cons of particular frameworks.

Carnap's 'easy' approach has recently regained popularity, in particular following Thomasson's (2015) detailed development of it, which includes answers to some stock objections. Among many things, Thomasson does away with Carnap's explicit appeal to frameworks, and the distinction between empirical issues and choices of the framework that was attacked by Quine. For Thomasson, it can be determined whether *K*s (say, smartphones) exist by checking whether the application conditions for the term 'K' are fulfilled. Application conditions are basic semantic rules of use that allow competent speakers of a language to evaluate situations as ones in which the terms would or would not refer (2015, p. 90). Since in

[2] This is not to say that it is impossible to defend the existence of (some, or many) artefacts in a neo-Aristotelean approach, just as neo-Quineanism is not necessarily incompatible with the exjstence of all artefacts.

our everyday lives, there are many situations in which we have no trouble determining whether a smartphone is involved or not, 'smartphone' has application conditions—and there is, on the easy approach, no more to the existence of smartphones than that. Again, no more than basic linguistic competence and empirical evidence is needed to answer existence questions. Even the Quinean quest for identity criteria can be deflated: rather than govern the existence of objects, such criteria concern the co-application of sortal terms in statements such as "*x* is the same *S* as that" (2015, Sect. 6.3 and Thomasson, 2007), the truth of which can again be established by competent language use and empirical evidence.

Clearly, this retains analytic ontology's focus on language use, but without the need for identity criteria stricter than co-application conditions. Without Carnap's appeal to linguistic frameworks, however, it is an open question which and whose language use provides the basis for ontology. Here, Thomasson leaves room for deferring to competent users and experts, but also, interestingly, seeks to revive Carnap's idea of external questions as practical decisions. One suggestion, using a term coined by David Plunkett and Tim Sundell (2013), is to consider such questions as part of *metalinguistic negotiation* (Thomasson, 2017): disputes on the way in which we should use language, or whether we should use particular terms or sets of interconnected concepts.[3] Such disputes are genuine rather than merely verbal, but they may persist even if disputants would agree about other facts or about how language is actually used.

Much of this is geared toward showing that there may be something substantive about the discussions between, say, Quineans and neo-Aristoteleans. It leaves unclear whether and how such debates could be resolved, and why one should care about the outcome unless one has vested interests in analytic ontology. Negotiations, metalinguistic or otherwise, crucially revolve around stakes. But many disputes on identity criteria—or the lack thereof—are based largely on intuitions and thought experiments, and any practical repercussions or negotiated stakes are at best implicit.[4]

[3] There are close connections to work on 'conceptual engineering' (see Thomasson, 2015, Section C.2).

[4] This does not hold for the stock examples of conceptual engineering—concepts such as 'race' and 'gender'. However, reconstructions of debates on those concepts as metalinguistic negotiation cannot be extended to debates on the existence of artefacts unless we identify the stakes.

There may be ways to identify what is at stake in the existing debate concerning the existence of artefacts, such that different ontologies can be taken as conflicting proposals to solve a practically relevant problem by conceptual means. Alternatively, we might take a practice-based route. This would start by identifying bodies of language use that are connected to societally vital, perhaps even controversial practices—so that it is clear from the outset that there may be something at stake. In such practices, we can then identify application and co-application conditions, or more interestingly, debates and decisions concerning such conditions, to arrive at one or more *practice-based* ontologies. My aim in the remainder of the paper is to construct such an ontology for technical artefacts. This would, then, require first of all the identification of language use, associated with a societally vital activity, that involves statements such as "This is an iPhone X" or "This device is the same smartphone as that one"—where, again, it should be clear what is at stake at accepting or denying, for instance, the last statement.[5] Without claiming that it is the only one, the next section presents one such body of language.

2 A Suitable Framework for Technical Artefacts: Patent Law

In the previous section, a practice-based variant of the 'easy' meta-ontology was proposed. On this approach, developing an ontology of artefacts requires, as a first step, identifying a suitable practice.

Here, we focus on legal practices concerning patents. Patents, and the activities associated with them (e.g., applying for, protecting, infringing), play a central role in contemporary society: Apple and Samsung alone have fought dozens of multi-million-dollar lawsuits citing patent infringements over the last decades. More generally, the practice has strong advocates, who consider patents to be essential in incentivizing innovation, which is in turn held to be essential to continued economic prosperity

[5] It is easy to think of contexts where such statements are meaningfully made about individual smartphones ("Hey, that is my phone!"), less straightforward for statements about kinds of phones.

and sustainable human well-being; but there are also outspoken critics, who maintain that patents stifle innovation. Moreover, artefacts are obviously central to patent law, which is also not obviously restricted to a particular subset of technical artefacts, unlike, for instance, theories in the engineering sciences. Finally, this practice comes with its own, specialized and circumscribed bodies of language, namely that used in bodies of law, written jurisprudence, reports on patent applications, and the applications themselves. Thus, we may expect to find in patent practice a wealth of statements regarding the identity and persistence of artefacts; and given the controversies surrounding the practice, any disputes on or conflicts between such statements might well be taken as signals of metalinguistic negotiation. In the context of Apple Inc. v. Samsung Electronics Co., a lot is at stake when making statements such as "This is the same graphical user interface as that".

There are important differences between legal systems regarding patents, which are partly coordinated by international treaties. Consequently, there is not a single unified practice associated with patents, and ontologies based on patent practices are necessarily 'local' ontologies. This section focuses on the European Union, in particular the 16th edition of the European Patent Convention (2016),[6] and the activities of applying for and granting a patent within this legal context. Still, no single section can do justice to every or even any aspect of this restricted practice, or its full linguistic complexity. Rather, I only outline some of the basic elements of patent law (2.1), before reviewing—with an equally broad brush—some aspects of particular interest (2.2).

2.1 Patents: Basic Elements

Patent law is one part of intellectual property (IP) law. Like all branches of IP law, it is meant to grant persons certain rights over their intellectual creations. Specifically, a patent owner has the right to exclude others from making, using, selling, or otherwise exploiting the invention for which a patent is granted, until the patent expires and provided that the invention

[6] Available at https://www.epo.org/law-practice/legal-texts/epc.html

is publicly disclosed. An invention, which may be a product or process, needs to satisfy the following three basic conditions (with references to the relevant articles of the EPC) in order to be patentable:

1. *Novelty*. An invention is not part of the 'state of the art', which is everything made publicly available before filing the patent. (Article 54)
2. *Inventive step*. The invention must not be obvious to a person 'skilled in the art' (i.e., someone who has the usual skills and knowledge in a technical field). (Article 56).
3. *Industrial applicability*. The invention can be made or used in any kind of industry, including agriculture (Article 57).

These conditions must make sure that only those ideas are protected which are sufficiently different from what is already known, and which can actually be realized. In practice, each condition is in need of further specification and has been gradually fine-grained in various documented decisions, also because of technological and social dynamics. The second condition, for instance, is acknowledged to involve a hypothetical entity (a 'legal fiction') that leaves considerable room for interpretation. The third condition, discussed in some more detail in Sect. 2.2, originally ruled out primarily inventions that were physically impossible, but it has also been used against patenting what would be applied only in the private sphere, such as certain methods of contraception (Decision T 74/93).

Filing a patent application requires a *specification* of the invention and a set of *claims*.[7] The specification needs to present the invention in relation to the prior art in a technical field, for example, mechanics, and as a solution to a technical problem, for example, stepping the pedals of a bicycle (Rule 42).[8] It must be presented in such detail that a person skilled in the art (the same fictional entity featured in the 'Inventive step' condition) can understand its operation, so that by the specification, the invention is effectively disclosed. Moreover, at least one way of 'carrying out' or 'embodying' the invention must be described. As part of the

[7] See https://www.epo.org/applying/european/Guide-for-applicants/html/e/ga_aii_2.html for an example.

[8] https://www.epo.org/law-practice/legal-texts/html/epc/2016/e/r42.html

specification, most applications include diagrams, often representing the prior art (conventional solutions to the technical problem) as well as the invention. The *claims* in a patent application specify the features of the invention for which protection is sought (EPC Rule 43). This must detail, on the one hand, how the invention draws on the prior art, that is, which features essential to it are already publicly available and, on the other hand, how it diverges from it, that is, which features distinguish it. Patent applications are thus by their format *cumulative* and *contrastive*.[9] They require describing how, on the one hand, an invention builds on the current state of the art and, on the other hand, diverges from it. Moreover, the format requires positioning an invention in a technological field, as a solution to at least one given problem. This reflects broadly shared insights on technological change, namely that it does not happen in a vacuum (i.e., disengaged from practical contexts) or *ex nihilo* (i.e., disengaged from existing technologies).

Patent applications are still more firmly embedded in a societal context because filing an application is a legal action: it requests certain rights and associated legal protection. Therefore, applicants are strongly advised to have their applications prepared by a patent lawyer, rather than filling out the forms themselves. Once the application has been filed, the European Patent Office (EPO) examines it in two stages (European Patent Guide, Chap. 5). In the first stage, besides some formal details, a first search is made of the prior art in the indicated field, resulting in a report that evaluates the preliminary patentability of the invention in terms of the basic conditions listed above. Based on this report, the application may be withdrawn or is subjected to a substantive examination in a second stage. This examination involves technically qualified examiners as well as, if necessary, a legally qualified examiner. Even if a patent is granted based on this examination, it may be opposed by other parties, challenged in lawsuits, or simply infringed.

[9] For this reason, patent databases have been used extensively to study processes of technological change.

2.2 Specific Features of Patentable Inventions

Based on the all-too-brief sketch of patent law just outlined, some features can be highlighted that are of particular interest for an ontology of technical artefacts. Two aspects in particular bear directly on the issue what counts as an invention and what distinguishes inventions from other kinds of objects and from each other; and both of them illustrate what is, in this practice, at stake in resolving such issues.

a. The boundaries of patent law

As mentioned above, patent law is only one part of IP law. The rights over other 'creations of the mind' are governed by closely related, but distinct bodies of law. *Copyright* law, for instance, governs a bundle of rights vested to an *author* who created a *work* such as a musical composition or a poem. Legally, some objects are eligible to be works, and others eligible to be inventions—and the distinction has important repercussions for what may be done with it. A work may, for instance, not be publicly displayed without permission of the copyright owner; although ownership of the copy of work also transfers some of these rights (think of playing a movie in a public square versus in one's home for friends). An invention, by contrast, has to be publicly disclosed for a patent to be granted and its patent owner has the right to (try to) *exclude* others from making or even using the invention.

There are paradigmatic examples of both works and inventions; say, poems and pedaling mechanisms respectively. Yet in practice the boundary between the two is fluid and heavily contested, and they are intertwined in subtle ways. To give an example of the latter: inventions such as pedaling mechanisms are governed by patent law, but in many legal systems the content of patent applications may be considered a work that is in principle subject to copyright, even if it is typically exempted from protection. As this makes clear, an invention is legally distinct from its specification in a patent; and likewise, a work is often distinguished from the ideas expressed in that work (e.g., Kant's transcendental idealism versus his *Critique of Pure Reason*), over which the work's author typically has no rights.

Turning to the fluidity of and contest over the boundary between inventions and other objects: there are broad categories of objects that cannot be recognized as inventions despite their obvious commercial or industrial importance.[10] One example besides works are industrial designs. For these ornamental designs of functional objects, such as the characteristic shape of the Coca-Cola bottle, protection can be granted, but by a different procedure, through a *registered design* (in the European Union), and without demonstrating an inventive step.

The legal distinctions between (functional) inventions and (ornamental) designs or between (technological) inventions and (creative) works put some objects—or rather those dealing with them in legal and economic practice—in an awkward place. A particularly complicated and increasingly important contested area concerns software, or any kind of digital object (Sterckx & Cockbain, 2012, Chap. 3). Article 52 of the European Patent Convention explicitly excludes computer programs as patentable inventions, along with "aesthetic creations" and "schemes, rules and methods for performing mental acts, playing games or doing business", while simultaneously restricting this exclusion to these items "as such". Far from settling the debate on the patentability of software, invoking this so-called 'business method exclusion' has in this case led to confusion, fierce legal battles, and the rejection in 2005 of a proposed EU directive on the patentability of 'computer-implemented inventions'. Originally, a distinction was made between a technical solution to a problem, and a non-technical one (e.g., in German law) or, because there are frequently software and hardware solutions to the exact same problem, between 'technical' and 'non-technical' software. Increasingly, however, EPO's Technical Board of Appeal recognizes software as invention if it presents a novel solution to a technical problem and contains an inventive step.

b. The Industrial applicability condition

[10] To be distinguished from objects that would be patentable, but are not patented precisely because of their importance. These include *trade secrets*, such as the formula of Coca-Cola.

The patent system was originally devised to advance social and economic progress through practically useful inventions (Pila, 2010, p. 8; Wadlow, 2008). The industrial applicability condition reflects this public policy ideal. It should ensure that legal protection or monopolies only extend to inventions that contribute to some practical purpose, or that are at least very likely to do so.

It is much disputed how 'industrial' should be interpreted to have this "gating function", and how detailed and concrete the indication of practical usefulness of a purported invention should be (Díaz Pozo, 2017, Sect. 3.4). EPC Article 57 suggests a broad interpretation, by explicitly mentioning agriculture as a field of activity. Also in practice, the condition is held to include any practical activity of a broadly technical character, not necessarily mechanical or electrical, much like its counterpart 'utility' in the US legal system. This might seem to exclude hardly anything apart from inventions contrary to the laws of physics, making the condition a "low hurdle for patent applicants to clear" (Díaz Pozo, 2017, p. 97), Still, that industrial application is a requirement for patentability means that a positive case must be made for applicability when filing a patent.

Recent rulings on biotechnology, genomics and medical treatments show that industrial applicability can be demonstrated on the basis of potential use; no evidence is needed for actual use at the time of filing or the very near future, let alone of commercially viable use. Yet this potential use must be described reasonably specifically; the 'person skilled in the art' should not need to guess what the use might be. Moreover, it does not suffice that something can merely be made or used. Rather, the use of an invention must consist in a contribution to producing a practically meaningful, tangible result in the context of a non-mental activity. Through the business method exclusion, the patent system should not facilitate monopolies over non-technical items such as procedures, treatments, or rules. In addition, however, inventions that were presented as having only research use were ruled to lack industrial applicability.

3 A Practice-Based Ontology: Artefacts as Abstract Types

The previous section outlined some basic elements and specific aspects of (a) patent law and practice. In this section, this is used as a basis for constructing an ontology of technical artefacts.

First, that artefacts exist can be derived directly from patent practice as described. Arguably once an application with the specification of an invention has been filed, but definitely once it has been granted, there are clear application conditions for a term referring to the invention. Here, the competent speakers that can evaluate whether the term does or does not refer in particular situations are the experts in the field ('persons skilled in the art') as well as legal experts. Moreover, there are co-application conditions that govern statements that inventions are the same or different: establishing this is pivotal to patent practice. Of course, these conditions may be contested, but this seldom takes the form of denying the existence of a particular type of invention. Rather, it may be claimed that the invention does not constitute a new type, but rather re-instantiates an existing one; or that what is specified in a patent application is not a patentable invention, but rather some other type of object (e.g., a work).

Doubts about the existence of technical artefacts can thus be side-stepped by casting them as patented inventions and referring to patent practice. This leads to a second issue: once we acknowledge the existence of artefacts, what are they?

Inventions as described above cannot be equated to any material object or class of such objects. A patent application necessarily lists a set of features, possibly including a schematic diagram, but there might, strictly speaking, not be (yet) any object that has this specific set of features even if the patent is granted. Moreover, the invention over which the patent owner has rights cannot coincide with any or all of the material instantiations of this invention, which can be owned by others: you have certain rights over the pedals over your bicycle, because you purchased that bicycle, because this has no implications for the rights that the patent owner has over the bicycle-pedaling invention. A practice-based ontology needs to account for this distinction, and placing the invention in an entirely different ontological category from material instantiations is a (conceptually) straightforward way to do so. Furthermore, the way in which the conditions are specified in the European Patent Convention also precludes that inventions are mental objects: patentability requires others to be able to recognize on the basis of the specification that the invention presents a novel solution to a technical problem, in much the same way as the inventor recognized this.

These considerations align with analyses of art works as abstract objects. In particular, several authors have argued for this conclusion on the basis of copyright or copying practices (see, e.g., Hick, 2011; Bahr, 2016; Reicher, 2016) or intellectual property more generally (e.g., Wilson, 2010). Despite salient differences, patent and copyright laws are embedded in one overriding legal framework that is meant to govern rights over various kinds of 'creations of the mind'. Moreover, the boundary between these kinds is dynamic, contested and to some extent permeable (recall the dispute on computer programs). A practice-based ontology should accommodate these continuities; accepting the arguments for works, casting works and inventions in the same ontological category is again a straightforward way to do so.

More in particular, we may extend to technical artefacts the *type ontology* that has been defended by the authors mentioned above, following less immediately practice-based proposals for aesthetic works (e.g., Currie, 1989; Levinson, 1980). This ontology construes the relation between inventions and its occurrences or instantiations as one between a type and its tokens. The identity of types is determined by the conditions that items must meet to be one of its tokens, and not by what are actually its tokens. This captures how inventions are distinct from and in many senses prior to actual devices instantiating them, and also how inventions are individuated according to patent laws: Sect. 2 has outlined how patent applications need to describe inventions in terms of their distinctive features, which set the type-determining conditions. Arguably, that there is a practice that requires artefacts to be specified in accordance with the definition of types makes a type ontology more easily applicable to technical artefacts than to, say, musical works.

Several authors have drawn attention to distinctive characteristics of intellectual creations as abstract types. For one thing, existing-type ontologies of works emphasize the relation between some mental processes in the process of creation and the created work. This process is for instance analyzed as consisting "essentially in *determining the essential features of realizations* of the work" (Reicher, 2016, p. 67; original emphasis), as "indicating" or "picking out" an abstract structure or type (Levinson, 1980, p. 82), as "sanctioning" with which a work is "endowed with certain features" (Irvin, 2005, p. 316) or more directly as "creating" an abstract design plan (Bahr, 2016, p. 91) or a new work (Hick, 2011). In all cases, a central role is reserved in the analysis for a mental act of creation, indication, or sanction that involves determining a set of type-determining properties. Most authors avoid further specification of these properties. This leaves sufficient room for adapting their construals to particular creative practices, including those governed by patent law.

A practice-based ontology of technical artefacts should accommodate the discontinuities as well as the continuities between patentable inventions and other 'creations of the mind'. Although it may be contested which created items count as patentable inventions, the very fact that this is contested shows that, legally, there is a real distinction to be made. Based on the discussion in Sect. 2, patent practice may lead to a specific, non-standard type ontology for technical artefacts.

One specific characteristic is that artefact types as patented inventions are, so to say, *epistemically transparent*. The type-determining features of technical artefacts are made fully explicit in a patent application: the specification includes those features that distinguish a candidate invention from the prior art, as well as shared features. It must also be established whether candidate inventions are sufficiently distinct from the prior art, that is, from existing types of technical artefacts. Originality in copyright law, by contrast, concerns whether a work may be regarded as an author's own creation; similarities to existing works only become relevant if those call into doubt whether someone has created the work through their own intellectual effort. Conversely, this effort is at best only implicit in the inventive step condition: the candidate invention should not be obvious to anyone skilled in the art. Thus, the exact conditions for 'indicating' an abstract structure (rather than, say, re-identifying it inspired by someone else's truly creative intellectual effort) or what is involved in an author's 'sanction' are beside the point: what matters to bring into existence a new type of technical artefact is whether the set of features described is as distinctive and non-obvious as claimed. For this, it is essential that *others* (versed in the art) can determine which objects would count as an instance of the candidate new type—and whether objects of existing types do, since this would undermine the claim to novelty. Consequently, for technical artefacts, identity criteria cannot be distinguished from individuation criteria: types of technical artefacts are determined by sets of features, and those with sufficient expertise and authority need to be able to differentiate artefacts on those grounds in practice.

This affects claims regarding the epistemic privilege of creators over others. Most prominently, Thomasson (2007) argues that, because creators' intentions are constitutive for the features relevant for artefact-kind membership (interpreted here as type membership), creators "are guaranteed certain forms of immunity from massive error about the objects of their creation" (2007, p. 64). Finding out that, say, pencils are actually organisms (Putnam, 1975) or alien listening devices (Nelson, 1982) would, according to Thomasson, only reveal that we are mistaken about which items are artefacts or who the creators are. The considerations above partly support this view: inventions need to have a set of characteristic features, and their creators must be mostly correct in attributing these features.[11] Yet this freedom from error should extend to others who review and grant the patent claim. Thus, it constitutes an epistemic *priority*, but it cannot be a *privilege* of the inventor over reviewers.

A second aspect of the distinction between works and inventions is based on the embedding of technical artefacts in broader human activities than patent practice. As discussed above, the appeal to the industrial applicability condition in recent rulings specified the relation between an invention and practical activities: in order to be industrially applicable,

[11] Thomasson also claims, in response to objections, that "the makers' concepts of what features are relevant to membership ... are typically superficial" (2007, p. 69). In a patent-based ontology, this claim seems doubtful: features described in a patent application typically go beyond "surface-level characteristics" (ibid.), for instance in specifying mechanisms. Thus, creators' epistemic access to technical artefacts is more extensive than Thomasson submits, even if it is less privileged.

an invention must produce some practically meaningful, tangible result in the context of a non-mental activity. Moreover, this potential practical use needs to be substantiated in a patent application. Consequently, there needs to be a clear indication of one or more activities to which the invention contributes (e.g., riding a bicycle) and the specification of its type-determining features must in combination with the prior art make clear *how* it contributes. It should thus not only be clear what the type-determining features of technical artefacts are, but patent law has requirements on what should minimally be included as type-determining features—namely a set of properties or dispositions through which the invention contributes to a practical activity. This means that a practice-based ontology of technical artefacts casts each artefact as a practically meaningful type, which is embedded in at least one broadly acknowledged practical activity.

One way to characterize these practically meaningful types is in *functional* terms. One reconstruction of function ascriptions—the ICE theory (Houkes & Vermaas, 2010)—defines those ascriptions in terms of (beliefs about) those capacities of artefacts that contribute to realizing the goal state of so-called use plans. In terms of this theory, the industrial applicability condition demands that inventors effectively ascribe *at least* one function to their creations in filing a patent. Interestingly, this suggests that artefact types are in part determined by their function. A reason to label technical artefacts as 'practically meaningful' rather than 'functional' ones is that the determining role of the function should not be overstated: it may amount to a first indication of a practical context in which the artefact potentially plays a useful role. Whether this suffices to claim, following Baker (2004) and several others, that artefacts have their functions *essentially* cannot be discussed in any detail here. Arguments against function essentialism revolve partly around scenarios of functional change or hybridization (e.g., Houkes & Vermaas, 2004). Incorporating those arguments into the practice-based ontology presented here would require finding relevant rulings on such scenarios.

4 Conclusion

This paper sketched an ontology of technical artefacts based on patent practice, more specifically on the European Patent Convention and recent rulings regarding, among other things, computer programs and genomics. In this context, artefacts were found to be understood as abstract types that are epistemically transparent to their creators as well as others, and that are practically meaningful as functional types. The latter further embeds the ontology in practice: legal practice is taken as the general basis for the ontology, and each individual artefact type refers to at least one practical activity.

In line with Thomasson's easy approach, no substantive argument has been given for the existence of artefacts, for instance to resolve the composition problem. Rather, it is taken for granted that patent practice provides clear application and co-application conditions for terms referring to specific artefact types. As noted, the latter conditions are often heavily contested in practice: statements that patentable inventions are the same thing are the subject of huge commercial and legal interests. More fundamentally, patent law and practice respond to societal change and pressures, for instance in their relation to other bodies of law and to concerns regarding open innovation. This, I submit, provides an example of metalinguistic negotiation—here, between inventors, patent lawyers, corporations, legislative bodies, and many others—in which forms of language use are challenged on the basis of conflicting societal stakes. A similar point may be made with regard to a concern about identifying artefacts as patented inventions: what about inventions that are not yet patented, or for which the patent claim has been rejected? I cannot address these concerns here, apart from noting that the practice-based approach entails that we look for answers in other societal practices than granted patent claims, and turn our attention to patent disputes, the free-culture and open-access movements that deliberately forego patent claims, or other more encompassing forms of language use and the metalinguistic negotiations triggered by them.

Neither the ontology of technical artefacts nor the practice-based meta-ontology have been developed in any depth in this chapter. Both are offered as proposals to put the ontology of artefacts and analytic ontology more generally on a different footing. This partly reflects growing concerns, both within the subdiscipline and more personally, that little or not is at stake in most debates within analytic ontology—whereas there are societally vital debates in which conceptual sophistication is readily applied and urgently needed.

I realize that none of this will and perhaps should convince committed Quineans that the ontology of technical artefacts or a practice-based approach is worth pursuing.[12] Some have, in response to concerns about the triviality of ontological debates, argued that such debates are waged in a specialist language called 'Ontologese' (e.g., Sider, 2011) or in an "ontology room" that is outside the "ordinary business of life" (Van Inwagen, 2014). This may indeed fit best how analytic ontology has developed in relation to other forms of language use. It also, however, amounts to a tactical withdrawal from metalinguistic negotiations with other language users. Yet something is to be said for this: it seems safest to reserve 'ontologese' for the ontology room rather than pit it against 'legalese' in the courtroom where patent lawsuits are resolved. In the ordinary business of life, when it would come to negotiations, it seems safest to bet one's money on the lawyers rather than the philosophers.[13]

References

Bahr, A. (2016). What Is an Artefact Copy?, In D. H. Hick & R. Schmücker (Eds.), *The Aesthetics and Ethics of Copying* (pp. 81–98). Bloomsbury Academic.

Baker, L. R. (2004). The Ontology of Artifacts. *Philosophical Explorations, 7*, 99–111.

[12] In this, my approach here is not an instance of what has been called "deferentialism", i.e., the idea that "authoritative disciplines outside philosophy" have a "decisive bearing on philosophical issues" (Daly & Liggins, 2011, p. 321). I do not mean to suggest that the bearing of patent practice will be that decisive: philosophers can and will dismiss any societal practice as insufficiently authoritative to be taken at face value or even to provide any substantial "evidence for philosophical debates" (ibid.). The relation between philosophical debates and societal and scientific practices is, in my opinion, even considerably more complicated than the 'anti-deferentialist' suggestion of settling on "the solution that offers the most benefits for the least costs" (ibid., p. 336).

[13] My sincere thanks to two anonymous reviewers for very helpful comments and suggestions to clarify or rephrase points in the chapter.

Berto, F., & Plebani, M. (2015). *Ontology and Metaontology*. Bloomsbury Academic.
Carnap, R. (1950). Empiricism, Semantics, and Ontology. *Revue Internationale de Philosophie, 4*, 20–40.
Currie, G. (1989). *An Ontology of Art*. St. Martin's Press.
Daly, C., & Liggins, D. (2011). Deferentialism. *Philosophical Studies, 156*, 321–337.
Díaz Pozo, M. (2017). *Patenting Genes*. Oxford University Press.
Hick, D. H. (2011). Toward an Ontology of Authored Works. *British Journal of Aesthetics, 51*, 181–199.
Houkes, W., & Vermaas, P. E. (2004). Actions versus Functions. *The Monist, 87*, 52–71.
Houkes, W., & Vermaas, P. E. (2010). *Technical Functions*. Springer.
Houkes, W., & Vermaas, P. E. (2014). On What Is Made. In M. Franssen, P. Kroes, T. A. C. Reydon & P. E. Vermaas (Eds.), *Artefact Kinds* (pp. 167–190). Springer.
Irvin, S. (2005). The Artist's Sanction in Contemporary Art. *Journal of Aesthetics and Art Criticism, 63*, 315–326.
Levinson, J. (1980). What a Musical Work Is. *The Journal of Philosophy, 70*, 5–28.
Lewis, D. K. (1986). *On the Plurality of Worlds*. Blackwell.
Lowe, E. J. (1998). *The Possibility of Metaphysics*. Oxford University Press.
Lowe, E. J. (2014). How Real Are Artefacts and Artefact Kinds? In M. Franssen, P. Kroes, T. A. C. Reydon & P. E. Vermaas (Eds.), *Artefact Kinds* (pp. 17–26). Springer. https://doi.org/10.1007/978-3-319-00801-1_2
Nelson, J. (1982). Schwartz on Reference. *Southern Journal of Philosophy, 20*, 359–365.
Pila, J. (2010). *The Requirement for an Invention in Patent Law*. Oxford University Press.
Plunkett, D., & Sundell, T. (2013). Disagreement and the Semantics of Normative and Evaluative Terms. *Philosopher's Imprint, 13*, 1–37.
Putnam, H. (1975). The Meaning of 'Meaning'. In *Mind, Language, and Reality* (pp. 215–271). Cambridge University Press.
Quine, W. V. O. (1948/1953). On What There Is. In *From a Logical Point of View* (pp. 1–19). Harvard University Press.
Reicher, M. E. (2016). What Is the Object in Which Copyright Can Subsist?. In D. H. Hick & R. Schmücker (Eds.), *The Aesthetics and Ethics of Copying* (pp. 61–79). Bloomsbury Academic.
Renz, G. (2016). It's All in Your Head. *Philosophia, 44*, 1387–1407.

Rosen, G., & Dorr, C. (2002). Composition as a Fiction. In G. Gale (Ed.), *The Blackwell Guide to Metaphysics* (pp. 151–174). Blackwell.
Sider, T. (2011). *Writing the Book of the World.* Oxford University Press.
Sterckx, S., & Cockbain, J. (2012). *Exclusions from Patentability.* Cambridge University Press.
Thomasson, A. L. (2007). Artifacts and Human Concepts. In S. Laurence & E. Margolis (Eds.), *Creations of the Mind* (pp. 52–73). Cambridge University Press.
Thomasson, A. L. (2015). *Ontology Made Easy.* Oxford University Press.
Thomasson, A. L. (2017). Metaphysical Disputes and Metalinguistic Negotiation. *Analytic Philosophy, 58*, 1–28.
Van Inwagen, P. (1990). *Material Beings.* Cornell University Press.
Van Inwagen, P. (2014). Introduction: Inside and Outside the Ontology Room. In *Existence Essays in Ontology* (pp. 1–14). Cambridge University Press.
Varzi, A. (2007). From Language to Ontology: Beware of the Traps. In M. Aurnague, M. Hickmann, & L. Vieu (Eds.), *The Categorization of Spatial Entities in Language and Cognition* (pp. 269–284). John Benjamin.
Wadlow, C. (2008). Utility and Industrial Applicability. In T. Takenaka (Ed.), *Patent Law and Theory* (pp. 355–382). Edward Elgar.
Wiggins, D. (2001). *Sameness and Substance Renewed.* Cambridge University Press.
Wilson, J. (2010). Ontology and the Regulation of Intellectual Property. *Monist, 93*, 450–463.

The Cyberspace Strikes Back: An Ontological Account of Social Networks

Enrico Terrone

In the late Sixties, Danish artists Susanne Ussing and Carsten Hoff made a series of works of contemporary art under the assumed name "Atelier Cyberspace". This is arguably the first occurrence of the word "cyberspace" ever. Still, Ussing and Hoff's artworks do not involve computer-generated spaces but rather "sensory spaces" consisting of "abstract, furniture-like sculptures and elements" (Lillemose & Kryger, 2015). The use of the term "cyberspace" to designate computer-generated space originates, instead, from William Gibson's (1982) short story *Burning Chrome*. Gibson resumes the term in his 1984 novel *Neuromancer*, characterizing its meaning as follows: "Cyberspace. A consensual hallucination experienced daily by billions of legitimate operators, in every nation, by children being taught mathematical concepts… A graphic representation of data abstracted from the banks of every computer in the human system. Unthinkable complexity. Lines of light ranged in the nonspace of the mind, clusters and constellations of data. Like city lights receding…" (1984, p. 67).

E. Terrone (✉)
University of Genoa, Genoa, Italy
e-mail: enrico.terrone@unige.it

E. Terrone, V. Tripodi (eds.), *Being and Value in Technology*,
https://doi.org/10.1007/978-3-030-88793-3_2

During the Nineties, the term "Cyberspace" became popular as a colorful way of designating the Internet or the Web. Yet, after a few successful years, the term seemed to be heading towards decline. Gibson himself, in 1996, expressed disaffection with the use of the term to designate computer-mediated phenomena, since the latter do not concern the boundary "between real and unreal" but rather that "between real and real" (in Kneale, 1996).

In his 2000 book *The Ontology of Cyberspace*, David Koepsell drew on Gibson's disaffection with his own lexical invention, developing a full-fledged philosophical account of computer-mediated phenomena. Koepsell argues that what one is tempted to call the cyberspace is nothing but a very complex and powerful technical artifact which, from an ontological perspective, do not differ from garden-variety artifacts. The so-called cyberspace, on closer inspection, is not a space but only a thing in space. The cyberspace thus reveals itself to be a myth which Koepsell aims to dispel.

I will argue that it is not so. That is, I will vindicate the notion of cyberspace, arguing that the Web, as well as social networks such as Facebook or Twitter, deserve to be subsumed under this notion since they have a special ontological status with respect to garden-variety technical artifacts. Specifically, in Sects. 1, 2, 3 and 4, I will show that garden-variety technical artifacts can belong to two distinct ontological categories, namely, concrete particulars and types. In Sects. 5, 6, 7 and 8, I will argue that social networks are neither concrete particulars nor types but rather abstract particulars, and this entitles us to cast them as sui generis spaces, namely, cyberspaces. In Sect. 9, I will differentiate the social networks as cyberspaces from the spaces generated by virtual reality systems. Finally, in Sect. 10, I will draw my conclusions, showing how Koepsell's skeptical challenge to the ontological relevance of the notion of cyberspace can be met.

1 What Is There?

A promising way of starting an inquiry into ontology consists in looking around. As Quine (1953, p. 1) puts it, "A curious thing about the ontological problem is its simplicity. It can be put in three Anglo-Saxon monosyllables: 'What is there?'". Ontology in the Quinean tradition, however, has answered this question in a quite austere manner, which relies on the natural sciences, thereby excluding human-made entities from the inventory of being. In this essay, instead, I will answer Quine's question in a non-Quinean manner, endorsing the descriptivist stance that Peter Strawson (1959, 1966) had developed drawing on Aristotle's *Metaphysics* and Kant's *Critique of Pure Reason*. While the Quinean ontology aims to carve nature at its joints (Sider, 2011), the Strawsonian ontology that I favor rather aims to carve culture at its joints, individuating categories of cultural entities that deserve to be cast as real kinds since they can offer compelling explanations of cultural phenomena and trace illuminating connections between them (cf. Millikan, 2000; Elder, 2004).

From this perspective, the ontological question can be answered relying on shared understandings that are rooted in perception and manifest themselves in linguistic uses. For instance, I can start answering the question by considering what there is in my room: one person, two cats, one table, four chairs, one lamp, one computer, six notebooks, four books, two pens, one pencil, and so on. This example suggests that organisms and artifacts are the main categories one deploys when one aims to answer the ontological question in everyday life. Both organisms and artifacts, in the first instance, can be cast as *particulars*, that is, entities that we can localize in the shared framework of our experience which involves space and time: "particulars have their place in the spatio-temporal system" (Strawson, 1959, p. 233). My two cats and I, as organisms, have our place in the spatiotemporal framework just as my table, my chairs, and my computer, as artifacts, have theirs.

2 Particulars, Universals, and Types

Particulars can be traced back to universals. As a particular, I can be traced back to the universal "Human being" just as my cats can be traced back to the universal "Cat", my table to the universal "Table", and my computer to the universal "Computer". Strawson characterizes a universal as "a principle of collection of like things" (1959, p. 226). The universal "Cat", for example, enables us to collect particular cats in virtue of the features they share.

Still, certain universals do more than just supplying principles of *collection* of like things: they also supply principles of *construction* of like things. Strawson (1959, p. 231) calls the latter principles "types", arguing that they are special individuals which, unlike particulars, do not have a unique place in space but enables us to construct "instances" or "tokens" which have their unique place in space. A paradigm case of types are words and sentences whose instances are particular utterances or inscriptions. The word "the", for example, is a type that is instantiated by three tokens (to wit, two uses and one mention) in the line I am writing.

Besides words and sentences, there are many other artifacts which have this ontological structure, as Strawson (1959, p. 231) points out: "The general title of 'types', often, though rather waveringly, confined to words and sentences, may well be extended. I have in mind, for example: works of art, such as musical and literary compositions, [...] makes of thing, e.g. makes of motor-car, such as the 1957 Cadillac, of which there are many particular instances but which is itself a non-particular; and more generally other things of which the instances are made or produced to a certain design, and which, or some of which, bear what one is strongly inclined to call a proper name, e.g. flags such as the Union Jack". (1959, p. 231)

Although types lack a distinctive spatiotemporal location and thus are not particulars, they differ from other universals since they tend to behave like particulars in two crucial respects. First, in subject-predicate sentences, types are normally designated by subjects rather than by predicates, as for example when one says: "The Union Jack is red, white, and blue". Second, types are often designated by demonstratives or proper names such as "Union Jack" or "1957 Cadillac". In the case of artworks, a sort of proper name is supplied by titles; in the case of sentences, by quotations.

As a universal, a type can function as a principle of collection of like things. Yet, unlike other universals, a type can do so in virtue of being the principle of construction of the collected things, namely its tokens. In this sense, as Strawson points out, the type is a sort of Platonic form whereby particulars can be produced: "The Platonic model [...]—an ideal form of which the instances are more or less exact or imperfect copies—is, in these cases, an appropriate model" (1959, p. 232). While in

Plato's ontology non-particulars determine the being of all particulars, in Strawson's variant this model only applies to artifacts: "The non-particulars here in question are all such that their instances are artefacts. But the concepts concerned are not just rather broadly functional, like those of other artefacts such as tables and beds. Rather, to produce an instance, one must conform more or less closely to more or less exact specifications. Fully to describe a non-particular of this kind is to *specify* a particular, with a high degree of precision and internal elaboration" (1959, p. 233).

Being the outcome of an act of specification made by human makers, types, unlike Platonic Forms, are not eternal. A type, as a principle of construction of like things, is a created entity which begins existing at a certain historical time. Types ultimately are norms for constructing things, and they have a historical existence just like other norms such as state laws or rules of games.

All particulars can be traced back to some relevant universal which enables one to *collect* them, but only certain particulars can be traced back to a type which had enabled one to *construct* them, specifying their features "with a high degree of precision and internal elaboration". On the one hand, particular knives can be grouped under the universal "Knife" even if those knives have unrelated histories of production. On the other hand, a type such as the 1957 Cadillac groups only those particular cars that were constructed by means of this type. All particulars counting as tokens of a type belong to the causal network that originates from the creation of the type itself as a principle of construction of its tokens: their histories of production are all branches of the same historical tree whose root is the creation of the artifact as a type.

3 Types in Art and Technology

In the ontology of art, the distinction between types and particulars can be exemplified by contrasting a symphony with a painting. The latter is a particular which has its place, say, in some gallery, while the former is a type which lacks a specific place of its own but can be instantiated by particulars tokens that are its performances (Wollheim, 1980; Wolterstorff,

1980; Dodd, 2007). One should visit the Prado to see Velazquez's *Las Meninas* but there is no specific place where one should go to listen to Mozart's *Jupiter Symphony*: whatever place in which this symphony is performed might be fine. Likewise, from an ontological perspective, novels can be cast as types to be instantiated by copies, and films as types to be instantiated by screenings.

As for technical artifacts, the ontological distinction between types and particulars tend to match the historical distinction between craft and mass production. While a handcrafted artifact is a unique concrete piece that was made by an artisan, a mass product is just one among many tokens which all derive from a type that was created by a designer or a team thereof (cf. Carroll, 1998; Evnine, 2016, 2019). Let me compare, for example, a table handcrafted by a carpenter and an Ikea "Nordviken" table. For sure, both these tables are concrete things, otherwise one cannot use them. Moreover, they both can be traced back to the universal "Table" as a principle of collection of like things. Yet, the handcrafted table is a fully-fledged particular whereas the Nordviken exists as a particular only in virtue of being the token of the Nordviken type, which specifies "with a high degree of precision and internal elaboration" how its tokens should be constructed.

Computer programs also can be cast as types to be instantiated by tokens which are complex events occurring in a computer. In computer science, these events whereby programs are instantiated are usually called "processes". For example, I am writing this essay by means of the computer program "MS-Word" which is a type instantiated by a process running on my computer. A computer instantiates a program through a process just like an orchestra instantiates a symphony through a performance. In both cases, the instantiation of the type is mediated by what Noël Carroll (1998) calls a "template", which in the latter case is the musical score and in the former, the digital code. The template mediates between the type and the token. Just as the musical score provides the orchestra with the proper instructions to instantiate the symphony, the digital code provides the computer with the proper instructions to instantiate the program.

Beside constituting the template that connects computer programs to the process that instantiate them, the digital code also affects the ontological profile of other artifacts, turning them into digital types, that is, types that specify in digital terms how their tokens are to be produced. Photographs and films, for example, were types to be instantiated by display-tokens long before the digital turn, and so were the sounds recorded by phonographic means (cf. Benjamin, 1936; Kittler, 1986; Carroll, 1998). Yet, the digital code provides us with a new, notational way of connecting images as types, or sounds as types, to their perceptible instances, thereby making instantiation much easier and cheaper to achieve (Terrone, 2014, 2018). Likewise, the digital notation can be exploited to turn industrial artifacts into digital types to be instantiated by 3D printers.

4 Technological Progress and Typification

The growing spread of the digital notation might suggest that the technological progress tends to turn any artifact into a type which can be specified with a higher and higher degree of precision and internal elaboration. First, the industrial turn from craft to mass production had moved artifacts from the domain of particulars to that of types. Then, the digital turn has enabled us to specify and instantiate types in a more fine-grained and effective way by means of the digital notation. Technological progress thus would go hand in hand with typification.

However, such a linear story can be questioned by considering cases in which digital technologies, combined with robotics, lead to production processes which are sensitive to the materials and the context of production, ending up in artifacts which are no longer types but rather particulars. Digital technologies, in this way, can overcome mass production, pointing towards a new sort of craftmanship in which robots may play the role once played by artisans (Carpo, 2011, 2020).

Another way of questioning the connection between digital technologies and typification consists in focusing on social networks such as Facebook, Twitter, Instagram, Tik Tok, and more generally on the World Wide Web. These networks surely are artifacts that are produced by means of digital technologies, and yet they can hardly be cast as types. I cannot run a token on Facebook or Twitter on my computer. Rather, I run a token of a browser to visit Facebook or Twitter. Likewise, I do not run a token of the Web on my computer but rather I run a token of a browser to visit the Web. For sure, Facebook, Twitter, and the Web can all be traced back to the universal "Computer Network", but this is just a principle of collection of similar things, not a principle of construction of similar tokens which involves a high degree of precision and internal elaboration. Hence, Facebook, Twitter, and the Web are not types. Yet, it would be odd to conclude that they are concrete particulars just like handcrafted tables or cupboards. So, what sort of entities are they?

5 Social Networks as Abstract Particulars

Facebook, Twitter, Instagram, Tik Tok, and the Web itself, I contend, are abstract particulars. On the one hand, these networks are abstract since they do not have a unique place in space, just like types. There is no unique place in the world where one should go to see Facebook, as well as there is no unique place in the world where one should go to listen to the *Jupiter Symphony*. On the other hand, these networks, unlike types, are particulars, since they do not have instances: they are not principles of construction of their instances. Facebook is not a principle of construction of "facebooks", Twitter is not a principle of construction of "twitters", and the Web is not a principle of construction of "webs". For sure, one can use Twitter to make twits, Facebook to make posts, and the Web to make webpages, but the latter things are just outcomes, not instances, of the former. The twits made by users do not manifest the internal structure and complexity of Twitter in the way the performances made by orchestras manifest its internal structure and complexity of a symphony as a type. One cannot multiply Twitter, or Facebook, or the Web, in the way in which one can multiply a type by producing new tokens. There is just one Twitter, one Facebook, one Web.

Particulars are individuated by their place in space while types are individuated by a principle of construction of like tokens. We can say that a certain particular exists since we can locate it in space, and we can say that a certain type exists since we can acknowledge that a certain principle of construction is in force within a social practice. But what about abstract particulars such as Twitter, Facebook, and the Web? What do their existence amount to?

6 Languages and Spaces

As artifacts, Facebook, Twitter, and the Web are created entities which have a beginning in time. The Web's existence, for example, begun on August 6, 1991; Facebook's on February 4, 2004; Twitter's on March 21, 2006. What changed in the inventory of the world after those dates? What sort of things entered this inventory?

A first term of comparison which may be helpful to understand the ontology of social networks is offered by languages. The Italian language, the Chinese language, the English language are not concrete particulars since they do not have unique places in space. Nor are they types since they do not enable us to construct things that manifest the internal structure and complexity of those languages. The Italian Language enables us to construct a variety of sentences but does not enable us to construct instances which manifest the whole Italian in the way in which a performance by an orchestra can manifest a whole symphony. All this suggests that languages are abstract particulars. They surely have a beginning in time, though as a matter of fact we might find it hard to establish when a certain language begun to exist (cf. Evnine 2016, p. 151).

Just as a language provides us with a system of elements and rules to make sentences, a social network provides us with a system of elements and rules to make communicative entities such as posts, twits, or webpages. In this sense, both languages and social networks are abstract particulars which favor communication. Yet, I contend, social networks and languages have different ontological relationships with the communicative entities they produce. On the one hand, it would be odd to identify a language such as Italian with the totality of utterances and inscriptions that express Italian sentences. On the other hand, it seems quite plausible to identify a social network with the totality of communicative entities it produces. To wit, the communicative entities produced by a social network exhibit a significant degree of unity and interconnection which cannot be found in the totality of utterances and inscriptions produced by means of a language. We can navigate a social network in a way in which we could not navigate a language. In this sense, it is worth comparing social networks to another sort of abstract particular: the Space.

Concrete particulars do have a place in space, but the Space itself does not have a place in space. Rather, the Space is what enables things to have a place in space. In this sense, the Space is abstract rather than concrete. Yet, the Space is not a type, since it does not count as a principle of construction on like spaces, unless one believes that the spatial universe is created by a God or a demiurge who might use it as a prototype to create other universes. At least from the human perspective, there is just one space. Being non-concrete and non-instantiable, the Space belongs to the ontological category of abstract particulars.

An interesting similarity between social networks and the Space can be drawn. They not only belong to the category of abstract particulars but can be also characterized as totalities of places. Just as the Space is the totality of physical places, Facebook is the totality of profiles and posts, Twitter is the totality of profiles and twits, and the Web is the totality of websites. Just like the Space, a social network, as a totality, cannot have instances. There cannot be a token of Facebook, besides Facebook itself, since the profiles and posts of the putative token should be part of the Facebook totality. The same holds true for other social networks and for the web itself. Here is the reason why social networks are not types but rather abstract particulars which unify a multiplicity of places into one manifold just as the Space does.

7 Cyberspaces

Social networks, as abstract particulars, resemble languages since they are artifacts which enable communication, but they also resemble the Space since they unify places. Borrowing Gibson's term, I will dub "cyberspace" the category to which such artificial spaces belong.

Although both the cyberspace and the Space can be cast as a totality of places, they differ in some important respects. First, cyberspaces differ from the Space because the former are artifacts created by human beings. A cyberspace has a beginning in time while the Space begins with time itself, forming a whole which one might call spacetime or universe. Despite this difference in origin, however, both cyberspaces and the Space change in time as their places change. Just like space and time unite to form a whole called spacetime or the universe, cyberspace and time unite to form a whole one might call cyberspacetime, that is, the historical development of a social network since its origin.

Cyberspaces differ from the Space also with respect to the way in which places are connected into a totality. The places of a social network are not part of a continuum, and thus they are not connected by physical distances. At most, they can be connected through links. In this sense, the cyberspace is a discrete totality of places whereas the Space is a continuous totality of places. Consequently, a place in the Space is identified by its location in a geometrical three-dimensional structure which corresponds to a Cartesian system of reference having real numbers as coordinates, while a place in a cyberspace is rather identified by a natural number, namely, its address. For example, a place in the web is identified by its URL (Uniform Resource Locator). On the one hand, places in a cyberspace can be visited by knowing their address, just like buildings in a city. On the other hand, one can immediately jump from one place of the cyberspace to another in a way one could not immediately jump from one place of the Space to another.

8 Experiencing the Cyberspace

The cyberspace differs from the Space as for the way in which places can be experienced by subjects. We experience the Space from the inside, in virtue of being creatures who have their place in that space. As Gareth Evans (1985, p. 369) points out—drawing on Strawson's (1966) reading of Kant—we conceive of the Space with ourselves "as located within it and tracing a continuous path through it." Conversely, we experience the cyberspace from the outside: we navigate it, but we do not have a place in it. Although we explore both the Space and the cyberspace, we are part of the Space in a way in which we are not part of the cyberspace. Our body traces "a continuous path" through the former, but not through the latter.

Making us part of the Space, our body is all we need to visit places in that space. By contrast, if we want to visit places in the cyberspace, we need some technical supplementation. Specifically, an interface is required enabling us to explore a space we do not inhabit. Browsers provide us with such an interface. Just as my body allows me to move in the Space and enjoy perspectives on its places, my browser enables me to move in the cyberspace and enjoy perspectives on its places. The difference is just that the body offers three-dimensional perspectives from the inside, namely the perceptual field, whereas the browser offers two-dimensional perspectives from the outside, namely the content of the browser's window.

In sum, the cyberspace is a space that we can explore but we cannot inhabit. At most, we can have usernames, profiles, or avatars which represent ourselves in the cyberspace as if they were our placeholders in that space, but we ourselves cannot be part of that space. Nevertheless, we can interact with the cyberspace just as we interact with the Space, gathering information from it or making changes in it.

That being the case, the cyberspace nicely fits with Cartesian dualism, to wit, the ontological view according to which the mind (*res cogitans*) interacts with things in space (*res extensa*) and yet does not belong to it. Although Cartesian dualism is no longer popular nowadays as an account of the relationship between the mind and the Space, it provides us with a helpful model of the relationship between subjects of experience and the cyberspace which they can explore and even modify from the outside, without belonging to it.

9 Comparing the Cyberspace with Virtual Reality

Virtual reality systems also can be cast as artifacts which enable us to experience a space different from ours. Both virtual reality and social networks exploit digital technologies to offer us a space which can be navigated as if it were a self-standing space even though, as a matter of fact, it rests upon hardware devices which have their place in our physical space (Chalmers, 2018). Yet, a crucial difference remains: virtual reality systems provide us with replicas of our continuous space whereas social networks provide us with a different kind of space, which is discrete rather than continuous.

Furthermore, virtual reality systems are nowadays produced and used as if they were types rather than abstract particulars. For example, Alejandro González Iñárritu's 2017 *Flesh and Sand* (*Carne y Arena*) is a virtual reality system which enables one to share the experience of immigrants at the border between Mexico and the United States. From an ontological perspective, *Flesh and Sand* is like a film which is instantiated each time it is screened. The only difference is that the instantiation of a virtual reality system such as *Flesh and Sand* does not occur on a screen which, in principle, is accessible to several spectators but rather on a headset, which is accessible to only one user. From an ontological perspective, virtual reality systems resemble video games, which can be cast as types whose instances are "playthroughs" rather than screenings (Kania, 2018, p. 187). Just like a player instantiates a videogame by playing it, a user instantiates a virtual reality system such as *Flesh and Sand* by using it.

In order to count as an abstract particular instead of as a type, a virtual reality system should be used by multiple users in a continuative manner. It is hard to find virtual reality systems that are used in this way nowadays, but the Matrix in the homonymous 1999 film by the Wachowskis is a good example, though a fictional one. The Matrix is a virtual space inhabited by all the human beings which spend their lives there as if it were their physical space. Just as there is only one space, which is the totality of physical places, there is only one Matrix, which is meant to be the totality of the virtual places generated by that system (Chalmers, 2005).

While a virtual reality system such as *Flesh and Sand* is a type which resembles films and videogames when considered from an ontological perspective, a virtual reality system such as the Matrix is better cast as an abstract particular, namely, a sort of artificial space which resembles cyberspaces such as social networks or the Web. Yet, a crucial difference remains as for the nature of the artificial space and the way in which it is experienced. The cyberspace, as I have characterized it in this essay, is a discrete space which is experienced from the outside whereas the virtual space of the Matrix, just like physical space, is a continuous space which is experienced from the inside, through a sort of perceptual illusion (which in the story of the film also involves a cognitive illusion, namely, the formation of wrong beliefs). An interesting intermediate case is the online environment *Second Life*, which appears as a continuous space just like the Matrix and yet, just like social networks, is experienced from the outside, by means of an avatar which can move through that artificial space.

10 Addressing Koepsell's Challenge

In his book *The Ontology of Cyberspace* (2000), Koepsell argues that the Web does not have any special ontological status. From an ontological point of view, the Web is just an artifact like many others. In this sense, the "ontology of the cyberspace" ultimately comes down to the claim that there is no cyberspace: there are just extremely powerful and flexible artifacts which have their place in our physical space. As Koepsell (2000, p. 124) puts it, "Expressions in cyberspace are just as real, and exist in space in the same way as expressions in other media. The term 'cyberspace' is misleading to the extent that it connotes a dimension apart from that of ordinary experience. Electrical charges take up ordinary space as atoms take up ordinary space". Hence, Koepsell (2000, p. 125) concludes, "Cyberspace is physical and all its components are physical" [...] Cyberspace is a medium composed of silicon chips, copper wires, magnetic tapes and disks, fiber optic cables, and every other component of computers, storage media and networks which stores, transmit and manipulate bits".

I agree with Koepsell that it is worth stressing that cyberspaces such as social networks or the Web depend on the physical elements that underlie them. Yet, assuming a descriptive approach in ontology, one is not forced to identify an entity with the material elements on which it depends. For example, a novel such as *Moby Dick* depends on ink and paper and yet it would be wrong to identify it with a certain amount of ink and paper, as shown by the fact that we now have also electronic instances of *Moby Dick*, which might keep existing also in a world in which there would be no ink and paper anymore. The nature of cultural entities such as books or cyberspaces is to be found in the way in which these objects are experienced and used rather than in the physical elements which underlie them. The essence of such cultural entities lies in their appearance rather than in their material substratum. Just like *Moby*

Dick can keep existing despite changes in its material bases, so a cyberspace can keep existing despite significant changes in the underlying technologies. A cyberspace is currently composed of silicon chips, copper wires, and all the other material constituents which Koepsell mentions and yet, in the future, it might be composed of different material constituents without ceasing to be the social network it is. This suggests that the nature of a cyberspace lies in how it presents itself to its users rather than in the physical elements that undergird such a presentation. As Chip Morningstar and F. Randall Farmer (2008, p. 2) put it, "a cyberspace is defined more by the interactions among the actors within it than by the technology with which it is implemented. [...] Cyberspace is necessarily a *multiple-participant environment*".

Once the cyberspace is properly distinguished from its material underpinnings, its ontological singularity can finally show up. The cyberspace is not a kind of artifacts like many others, only more sophisticated and complex. There is an ontological peculiarity that differentiates the cyberspace from the other artifactual kinds. Most artifacts are either concrete particulars, as in the craft tradition, or types, as in the industrial tradition. Cyberspaces, instead, are abstract particulars. They are abstract since they do not have a place in space, despite resting upon material components which have a place in space. And they are particulars since they cannot have instances: the putative instances of a cyberspace would, on closer inspection, reveal themselves to be to regions of that cyberspace as a totality of places.

The Web, Facebook, Twitter, and the other social networks exist in space in the sense that they rest upon material components. Moreover, people use such cyberspaces to talk about real people and events, and thus one might be inclined to agree with Koepsell's (2000, p. 130) skepticism about the cyberspace as "a new realm, a place apart from the world of ordinary experience". However, all this is just about the current, contingent matter and contents of the cyberspace. If, instead, we reflect on the essential form of the experience which social networks are meant to elicit from us, we can finally recognize them as peculiar spaces that we do not inhabit with our body and yet we can effectively explore and modify through our actions. The cyberspace replaces the space of physical bodies and distances with a new space tailor-made for our mind.

References

Benjamin, W. (1936/2008). The Work of Art in the Age of its Technological Reproducibility, English translation: in *The Work of Art in the Age of its Technological Reproducibility and Other Writings on Media*. Harvard University Press.

Carpo, M. (2011). *The Alphabet and the Algorithm*. MIT Press.

Carpo, M. (2020, March). Rise of the Machines. *Artforum*, pp. 172–180

Carroll, N. (1998). *A Philosophy of Mass Art*. Clarendon Press.

Chalmers, D. (2005). The Matrix as Metaphysics. In C. Grau (Ed.), *Philosophers Explore the Matrix* (pp. 132–176). Oxford University Press.

Chalmers, D. (2018). The Virtual and the Real. *Disputatio, 9*(4), 309–352.

Dodd, J. (2007). *Works of Music: An Essay in Ontology*. Oxford University Press.

Elder, C. (2004). *Real Natures and Familiar Objects*. MIT Press.

Evans, G. (1985). Molyneux's Question. In *Collected Papers* (pp. 364–399). Clarendon.

Evnine, S. (2016). *Making Objects and Events: A Hylomorphic Theory of Artifacts, Actions, and Organisms*. Oxford University Press.

Evnine, S. (2019). Mass Production. In B. Brewer & J. Cumpa (Eds.), *The Nature of Ordinary Objects* (pp. 198–222). Cambridge University Press.

Gibson, W. (1982, July). Burning Chrome. *Omni*.

Gibson, W. (1984). *Neuromancer*. Harper.

Kania, A. (2018). Why Gamers Are Not Performers. *Journal of Aesthetics and Art Criticism, 76*(2), 187–199.

Kittler, F. (1986/1999). *Gramophone Film Typewriter*. English Translation. Stanford University Press.

Kneale, D. (1996, September 26). From the Man Who Named Cyberspace, a New Trip. *Wall Street Journal*, Eastern Edition. https://www.wsj.com/articles/SB843620903987895000

Koepsell, D. R. (2000). *The Ontology of Cyberspace: Philosophy, Law, and the Future of Intellectual Property*. Open Court Publishing.

Lillemose, J., & Kryger, M. (2015). The (Re)invention of Cyberspace. Nordic Art Review, https://kunstkritikk.com/thereinvention-of-cyberspace/

Millikan, R. G. (2000). *On Clear and Confused Ideas: An Essay about Substance Concepts*. Cambridge University Press.

Morningstar, C., & Farmer, F. R. (2008). The Lessons of Lucasfilm's Habitat. *Journal for Virtual Worlds Research, 1*(1), 1–21.

Quine, W. V. O. (1953). *From a Logical Point of View*. Harvard University Press.

Sider, T. (2011). *Writing the Book of the World.* Oxford University Press.

Strawson, P. F. (1959). *Individuals.* Methuen.

Strawson, P. F. (1966). *The Bounds of Sense: An Essay on Kant's Critique of Pure Reason.* Methuen.

Terrone, E. (2014). The Digital Secret of the Moving Image. *Estetika: The Central European Journal of Aesthetics, LI/VII*(1), 21–41.

Terrone, E. (2018). Appearance and History: The Autographic/Allographic Distinction Revisited. *The British Journal of Aesthetics, 58*(1), 71–87.

Wollheim, R. (1980). *Art and its Objects: An Introduction to Aesthetics,* 2nd ed.. Revised. Cambridge University Press.

Wolterstorff, N. (1980). *Works and Worlds of Art.* Clarendon Press.

Cognitive Artifacts Between Cognitive Sciences and the Philosophy of Technology

Marco Fasoli

1 Introduction: Cognitive Artifacts and the Philosophy of Technology

The concept of cognitive artifacts was introduced at the beginning of the 1990s by cognitive scientists, with the intention of identifying a sub-set of artifacts that support human cognitive processes. According to Norman, for instance, cognitive artifacts are "artificial instruments that support, display, or process information in order to perform a representational function and that affect human cognitive performance" (Norman, 1991, p. 17). Think of a map that you employ for navigating an unknown city or a shopping list that you compile before going to the supermarket. Both of these support cognitive processes, namely orienteering and memorisation, and they do that by representing information. A map represents spatial information, while a shopping list represents a list of the products you need to buy.

M. Fasoli (✉)
Scuola Universitaria Superiore IUSS Pavia, Pavia, Italy
e-mail: marco.fasoli@iusspavia.it

E. Terrone, V. Tripodi (eds.), *Being and Value in Technology*,
https://doi.org/10.1007/978-3-030-88793-3_3

Unlike in the case of the extended mind theory (Clark & Chalmers, 1998), cognitive artifacts have not been the subject of intense philosophical debate, even though their postulation raises several philosophical questions. This is somewhat surprising, especially if we consider the fact that the most hotly debated argument of the extended mind theory involves a cognitive artifact, namely Otto's notebook. In their article, Clark and Chalmers compare the cognitive capacities of Inga and Otto, the latter of whom suffers from Alzheimer disease. When they address cognitive tasks, Inga exploits her cognitive abilities, while Otto needs the help of an external device. To remember the location of MOMA, for instance, Inga relies on her biological memory while Otto consults his notebook, which plays "for Otto the same role that memory plays for Inga" (Clark & Chalmers, 1998, p. 4). It is well-known that the thesis of Clark and Chalmers concerns the ontological nature of the mind, which they consider is extended into the physical world in some specific cases. However, the thesis also involves a specific role of technological artifacts, which have often been neglected.

One of the possible reasons for the difference in popularity between the debate on cognitive artifacts and that on the extended mind lies with their different nature. While the extended mind argument is genuinely philosophical as well as provocative, insofar as it leads us to question our intuitions about the nature of the mind, cognitive artifacts are probably not as evocative a research topic. Moreover, as we will see, investigating the nature of cognitive artifacts often requires moving back and forth between the philosophical domain and the domain of cognitive sciences. This is particularly true when attempting to understand how the cognitive advantages provided by the artifacts arise.

In this chapter I take stock of what has happened in the philosophy of cognitive artifacts over the last 10–15 years. I present the main philosophical problems surrounding these objects and discuss some of the most significant answers that have been provided by philosophers. The issues can basically be grouped into four categories regarding the definition of cognitive artifacts, their ontological characterisation, the development of a taxonomy of these objects and the ethical aspects surrounding them. On the consequences, this chapter starts with a brief review of the different definitions of cognitive artifacts in the literature and proceeds by focusing on the metaphysical characterisations that have been given to these objects. Then, the classifications of cognitive artifacts is analysed and, finally, some ethical issues are discussed.

2 What Is a Cognitive Artifact?

Following the definition of cognitive artifacts provided by Norman, Edwin Hutchins (1999) provided a different definition in his entry included in the MIT Encyclopaedia of the Cognitive Sciences dedicated to these objects. According to Hutchins, cognitive artifacts are "physical objects made by humans for the purpose of aiding, enhancing, or improving cognition" (Hutchins, 1999, p. 126). Both Norman and Hutchins were cognitive scientists involved in the "situated cognition" debate, and interested in understanding how these objects contribute to cognitive human performances. In philosophy, the first to develop a comprehensive account of cognitive artifacts was Richard Heersmink, with a paper dedicated to the taxonomy of these objects (Heersmink, 2013). Heersmink defined cognitive artifacts as objects that "contribute functionally to the performance of a cognitive task" (2013, p. 46). It is an articulate definition, but one that refers to some controversial notions. For instance, establishing when something contributes functionally to the performance of a task and when it does not is far from easy (Vaccari, 2017), mainly because of the theoretical difficulties in developing a compelling definition of function (see, e.g. Ariew et al., 2002). Subsequently, Fasoli (2018) has defined cognitive artifacts as those "physical objects that have been created or modified to contribute to the performance of a cognitive task, providing us with representations that replace our cognitive processes, are complementary to them or constitute a condition of their possibility, thus modifying the original cognitive task or creating a new one" (Fasoli, 2018, p. 681). This definition, like the definition developed by Norman (1991), emphasises the representational nature of cognitive artifacts.

Despite the central role of the notion of a "cognitive task" in these definitions of cognitive artifacts, it seems that neither philosophers nor cognitive scientists have been concerned with discussion of the concept (one of the few exceptions is Carroll, 1993). First and foremost, a task appears to relate to performance and to include a goal that may or may not be reached, within a given time frame (Fasoli, 2019). In this sense, it appears as a concept related to the experimental sphere and to the need to obtain a measurement. However, several activities would appear not to meet these requirements, but at the same time to involve both external objects and demanding cognitive processes. Consider a child freely playing with a basic Lego set. He performs several complex cognitive processes involving a material artifact, but in this case it is hard to identify a measurable and tangible goal. While a calculation performed with an abacus has an objectively correct, exact result, which can be identified or not, in many other circumstances there is simply no result to be obtained.

A second set of problems concerns the difficulties in drawing a distinction between the cognitive sphere and others such as the perceptual, (Fasoli, 2018) affective (Piredda, 2019; Heersmink, 2021) and motor spheres. For example, let us analyse the case of a basketball player engaged in a free throw. It is a task (that can therefore be executed correctly or not) and involves perceptive, motor and visual processes. Can we really exclude the possibility that it is—at least in part—a cognitive task? In general, if we focus on what have most often been considered prototypical cognitive artifacts, for example, maps, checklists, abacuses and so on, we can see how the literature on cognitive artifacts seems to have implicitly assumed a particular meaning of the term 'cognitive', namely that of high-level reasoning processes (Fasoli, 2018, p. 593), without having duly discussed this assumption.

A distinction worth mentioning in this context is the one between artifacts, natural objects and techniques (Heersmink, 2021). This because human beings enhance and support their cognition in various ways, and only some of these involve what are known as cognitive artifacts. Heersmink (2013), for example, maintains that humans sometimes use natural objects to carry out tasks. He discusses the case of Micronesian navigators (Hutchins, 1995, p. 172), who use the stars as perceptual anchors to orient themselves. In this case, the stars are not specially constructed artifacts for navigation, they are natural objects or "naturefacts" (Hilpinen, 2018). This example is used by Heersmink for introducing another distinction, namely the distinction between 'cognitive artifacts' and 'cognitive techniques', the latter being methods or procedures used for solving a cognitive task, without the necessary involvement of a material artifact (see also Heersmink, 2021). According to Heersmink, what Micronesian navigators do is to put into practice orienteering techniques that involve natural objects (the stars) but not cognitive artifacts. Other cognitive techniques that can be mentioned here are "mnemotechnics"

3 The Ontology of Cognitive Artifacts

With the possible exception of cognitive scientists or philosophers, people do not usually use the expression 'cognitive artifacts' in everyday language. This is why introducing the category constitutes a revision of our intuitions with respect to objects in the world. Among the types of artifacts, the proponents of cognitive artifacts believe that it is necessary and useful, at least in some domains, to recognise the existence of what we call cognitive artifacts. In this sense, the recognition of cognitive artifacts is a theoretical move that is part of conceptual engineering (Cappelen, 2018). To be more precise, it constitutes a real conceptual innovation (Burgess & Plunkett, 2013, p. 1094). For the same reason, cognitive artifacts may also be one of the most unwelcomed theoretical novelties for artifactual eliminativists (see, e.g. van Inwagen, 1990), according to whom artifacts have no ontological consistency and should be expunged from our catalogue of things (see also Preston, 2020). In fact, by supporting the need to introduce a new kind of artifact, cognitive artifact supporters do exactly the opposite.

Being a kind of "artifact", cognitive artifacts are clearly affected by the same ontological issues of artifacts in general. As already mentioned, one issue concerns the difficulty in accounting for the great diversity of functions that users may attribute to the same object. These difficulties pushed Vaccari (2017) to argue against cognitive artifacts, and specifically against Heersmink's account (2013, 2016), also lamenting the lack of utility of this category for the extended cognition theory. Drawing on the theory of technical function developed by Preston (1998), Heersmink (2016) distinguished between the proper function of cognitive artifacts and the system function. The first is ascribed on an aetiological basis through a process of cultural selection, while the second is an extemporary function. The proper function of compasses is to facilitate orienteering because they "have been selected for this purpose by previous generations of users" (Heersmink, 2016, p. 39). Whereas the system function of a cognitive artifact is the one that it performs at a given instant, and such a function may be "cognitive" or not. For instance, a compass can be used as a paperweight, in this case attributing to the object a non-cognitive "system function".

The most important issue regarding the metaphysics of cognitive artifacts concerns the explanation of their capacity to play a cognitive function, namely their capacity to support human cognitive processes. What is it about these objects that differentiates them from the other artifacts? According to Heersmink (2016), the cognitive function of cognitive artifacts is a kind of emergent property arising from their physical structure in the following way:

$$\text{Physical structure} \rightarrow \text{representational structure} \rightarrow \text{information} \rightarrow \text{cognitive function}$$

(Heersmink, 2016, p. 9)

Take a cognitive artifact such as a map. Its cognitive function emerges from the ability to use it to encode spatial information through an isomorphic image. Hence, the cognitive function of this artifact emerges from the specific information it encodes. Such information is relevant to a particular task and originates from a representational structure and a physical structure.

However, it has been noted that this account seems to ignore some important differences between cognitive artifacts. Take a medium-sized paper map and its PDF version (Fasoli, 2018, p. 597, see also Fasoli, 2019). In this case, the information that is represented by the two artifacts is the same, but they offer different cognitive advantages and slightly different functions. The PDF map may be enlarged or shrunk as needed, while the paper version has a scale that defines a fixed ratio between the size of the representation and that of the represented. Furthermore, when a user searches for the name of a street on the two maps, she carries out different operations. Whereas in the case of the paper map, she has to start a visual exploration in order to identify the name on the sheet of paper, in the PDF version she can simply enter the name of the street in the appropriate search box. A concept that is useful to explain these differences is the concept of affordance of an artifact, imported from the philosophy of biology. In the artifactual domain, the affordances of an artifact constitute the set of possible uses of an object with respect to a specific user (see Gibson, 1979; Heersmink, 2016; Cosentino, 2019). Considering the aforementioned case of the two versions of the same map, it is possible to notice an 'informational equivalence' between the two objects and a difference in terms of their affordances. Furthermore, different cognitive affordances imply different cognitive advantages (Casati, 2017). For example, the PDF map allows one to locate several places in a very short time using the "search" function.

In an attempt to account for this kind of differences, Fasoli (2018) proposed this integration to Heersmink's diagram:

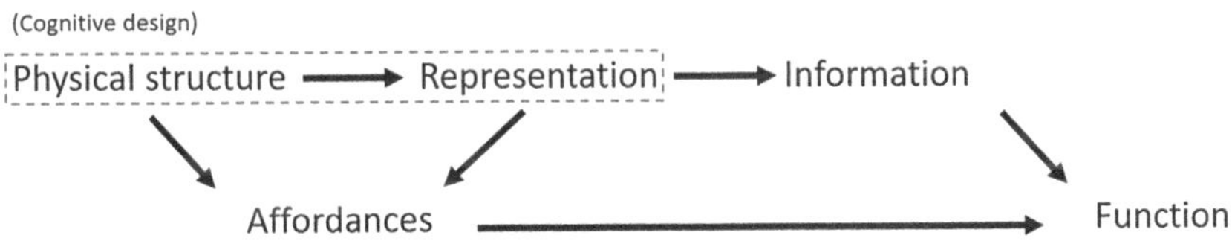

(Fasoli, 2018, p. 598)

Another way to explain the capacity of cognitive artifacts to fulfil their proper function is to focus on how artifacts can be used to transform the original cognitive task into another. The abacus, for instance, can be employed to transform the task "perform a complex mental calculation" into the task "use the abacus to create numerical representations, store them, break down a complex calculation into several simple calculations, and perform those simple calculations using the numerical representations". A GPS navigator transforms the task "memorise a route and follow it by identifying appropriate landmarks" into the task "input information in an appropriate way and follow the directions given". In this perspective, Casati (2017) has identified some of the "cognitive mechanics" that cognitive artifacts offer their users:

> freeing up working memory, storing information, making inferences visually available and available for inspection, priming action, bridging different modules, shunting information, providing hyperstimuli to a module, making orienting and navigation possible, indexing, displacing complex searches to the visual system. (Casati, 2017, p. 15)

These cognitive mechanics constitute the cognitive advantages that these objects provide to the users and that make them effective in supporting human cognitive processes.

4 Taxonomies of Cognitive Artifacts

In recent years, several scholars have studied the cognitive processes underlying the categorisation of artifacts (Bloom, 1998; Gelman & Bloom, 2000; Carrara & Mingardo, 2013; Malt & Sloman, 2007). One of the main goals of this research area is to understand on the basis of which properties people categorise different objects as belonging to the same type, for instance, how stools, rocking chairs, traditional chairs and so on are categorised as types of 'chair'. As far as cognitive artifacts are concerned, there are no studies on the categorisation of these objects, simply because (as mentioned in the previous section) humans do not usually categorise these artifacts. However, some scholars have developed taxonomies in order to distinguish different types of cognitive artifacts. The basic idea, in this case, is that a taxonomy of cognitive artifacts could be useful for understanding how these objects interact with our cognitive processes.

One of the first to attempt to taxonomise cognitive artifacts was Brey (2005), with an approach focused on the kind of cognitive capacity they support. For instance, a shopping list is considered a mnemonic cognitive artifact, a map an orienteering one and so on. While the taxonomical approach developed by Brey centres on human capacities, Heersmink (2013) has attempted to develop an "artifact-centred" approach. In other words, he has classified artifacts on the basis of their "intrinsic" features, independently from the kind of abilities they support. According to Heersmink, cognitive artifacts are a family that can be divided into two genera: representational and environmental (or non-representational). There are three species of representational artifacts, distinguished by Heersmink on the basis of Peirce (1935): iconic, indexical and symbolic. Finally, there are two species of environmental artifacts: spatial and structural (Fig. 1).

To introduce another taxonomy, Casati (2017) compares navigation in space using a GPS navigator with navigation using a map. In the former case the subject delegates the mental processes to the external object, whereas in the latter she has to employ a great amount of cognitive resources in order to successfully orient herself. These two different ways of using artifacts illustrate two modes of functioning of the mind, M3 and M4, which in many respects resemble the well-known distinction between S1 and S2, or better M1 and M2 in Casati's terminology. According to some scholars (see, e.g. Evans & Frankish, 2009; Kahneman, 2011), we should distinguish between two basic modes in which most humans' cognitive processes are realised, namely M1 and M2. Put very succinctly, the former operates quickly and automatically, "with little or no effort, and no sense of voluntary control" (Kahneman, 2011, p.110) while the latter is governed by logic and it is more precise, but it is also slow to activate. For instance, human cognitive biases arise from

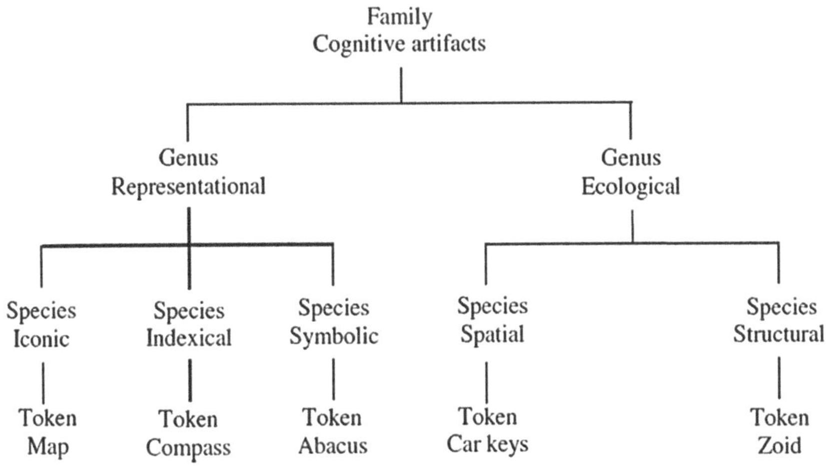

Fig. 1 The taxonomy developed by Heersmink (2013)

cognitive mechanisms that are usually realised in M1, while in most cases complex maths operations require M2 to be solved in the mind.

According to Casati, we should expand on this account by introducing two other fundamental modes of cognition, namely M3 and M4. In M4 mode, "the core computational task of navigation is clearly offloaded" (Casati, 2017, p. 12) onto an artifact, for example a GPS navigator. Consequently M4 mode roughly corresponds to M1, but with the involvement of a cognitive artifact. When operating in M4, a subject quickly accesses the input and the output of computation that is provided by an artifact, with a quite limited cognitive effort. On the other hand, M3 mode is characterised by the contribution of an artifact, such as a map, and by slowness and depth, similarly to what happens in M2. However, M3 is "distinct from M2 in that it essentially uses external artifacts" (Casati, 2017, p. 13).

Finally, Fasoli (2017) has developed an "interaction-based approach", taxonomising cognitive artifacts according to the type of relationships they have with our mental processes. In this context, he has distinguished between complementary, substitutive and innovative cognitive artifacts. Complementary cognitive artifacts complement our cognitive processes. For instance, a map needs orientation and interpretation in order to be used. It provides information that in some sense needs to be complemented by the subject's cognitive processes (see also Sutton, 2010). On the other hand, devices such as a GPS navigator system entirely replace the processes involved in spatial orientation, inasmuch the subject only needs to follow its instructions. Lastly, human beings have technologised their environment to such an extent that some artifacts are essential for carrying out certain cognitive tasks. Take reading, for instance. Written texts, whether they on paper, stone or any other material support, are essential for reading

From this overview, it emerges that cognitive artifacts are a heterogeneous class of objects that can be taxonomised in several ways. It is difficult to argue that one of these classification approaches is better than the others, possibly because their effectiveness also depends on the specific purpose for which the classification was designed.

5 Ethical Issues and Cognitive Artifacts

In several cases it appears plausible to consider the support provided by cognitive artifacts to our cognitive processes as a form of enhancement. For instance, calculators enhance our calculating abilities, compasses our orienteering capabilities and so on. However, cognitive enhancement by means of technological artifacts has not been the object of as intense a debate as psychopharmaceuticals. On the contrary, the moral aspects of cognitive artifacts have only recently begun to be addressed by few philosophers (see, e.g. Heersmink, 2015; Fasoli, 2016; Tigard, 2021).

It is possible to identify at least five general moral aspects related to cognitive artifacts. First, sometimes they can push people towards specific behaviours that are considered to be more proper and that engineers may decide to encourage by means of devices. For instance, by announcing that when a driver exceeds the speed limit, a GPS navigator will tell her to slow down, sparing her the cognitive effort of constantly checking her speed while driving and promoting moral behaviour (respecting speed limits). The second moral aspect arises as cognitive artifacts provide users with advantages that may be necessary and right to give assurance to everyone (Norman, 1993; Brey, 2005, p. 129; Heersmink, 2015). The "Digital Divide" (Di Maggio & Hargittai, 2001) can be considered a well-known example of this kind of moral issue, mainly because of the epistemic advantages the internet provides to users.

Thirdly, in as much as a cognitive artifact provides people with some cognitive advantages, it is necessary to determine in which contexts the use of such an object should be considered a form of cheating (see also Heersmink, 2015). Again, consider a calculator. For a child taking a maths test it could constitute a form of cheating, while for an engineer designing a bridge, it is simply an enhancing technology, one that saves a lot of time. In this case the difference between use for enhancement and use for cheating seems obvious, but there are other situations that are more complicated to evaluate. For instance, it is not easy to evaluate internet usage in educational contexts, especially considering the ease of access in our daily lives. Should we really still consider internet use a form of cheating nowadays, or we should grant it the status of "conventional" life-enhancing technology in every context?

The fourth moral aspect surrounding cognitive artifacts concerns the human brain. Under specific circumstances some cognitive artifacts may have unintended consequences on the cognitive system (Heersmink, 2015, pp. 8–9; Fasoli, 2016). They may become substitutes for other cognitive abilities that require frequent use for their retention, such as orienteering (Fasoli, 2016). A key problem, here, is to understand how to manage those new artifacts that provide users with some advantages but simultaneously may cause a reduction in some cognitive abilities.

The last moral aspect of cognitive artifacts concerns the effects that the technologies that record and store different aspects of user lives have on their personal identity. Heersmink (2015, 2017) discusses the case of Sense Cams, small portable cameras that are hung around the necks of some patients suffering from memory problems (e.g. due to diseases such as Alzheimer's), which automatically take photos at regular intervals. These mnemonic artifacts and the photos they create seem to play a decisive role in establishing and maintaining the personal identity of patients. A similar role, albeit to a very different degree, is played by social networks for many non-pathological subjects. Lifelogging technologies raise some of the aforementioned problems, such as the one regarding their effect on human mnemonic abilities (when they are used by non-pathological subjects). But they also raise new questions, for instance, regarding their effect on the narrative created by subjects and the possibility of inconsistencies with internally stored narratives (Heersmink, 2017).

In addition to these specific problems, of course, there are the moral issues that affect all technological artifacts, such as the problem of their moral value (Kroes & Verbeek, 2014; Pitt, 2014; Fasoli, 2019).

6 Conclusion

While it is widely acknowledged that technology is not a phenomenon specific to homo sapiens, but is also present in the animal domain (Preston, 2020), can we argue that cognitive artifacts are? This question may only be answered by biologists or ethologists. What is certain is that cognitive artifacts are a fundamental class of objects in our modern world, and that in the coming years we are likely to see the emergence of several new complex cognitive artifacts, based also on A.I. technology, for example in the decision-making domain. In this perspective, a deeper philosophical analysis of cognitive artifacts as well as of their ethical aspects will be of paramount importance. My hope is that this chapter may help in clarifying the most important aspects that have been addressed by philosophers so far and stimulate a renewed debate about them.

This is a structurally interdisciplinary field of research, insofar as, in addition to the argumentative and conceptual tools typical of analytical philosophy (in particular, the philosophy of technology), it requires knowledge of the cognitive sciences. Without an adequate mastery of the dynamics that govern our brains, it is not possible to understand the mechanics of the functioning of these objects, nor to hypothesise the effects they have in the medium to long term, from which, among other things, ethical problems arise.

References

Ariew, A., Cummins, R., & Perlman, M. (Eds.). (2002). *Functions: New Essays in the Philosophy of Psychology and Biology*. Oxford University Press.

Bloom, P. (1998). Theories of Artifact Categorization. *Cognition, 66*, 87–93.

Brey, P. (2005). The Epistemology and Ontology of Human-Computer Interaction. *Minds & Machines, 15*(3–4), 383–398.

Burgess, A., & Plunkett, D. (2013). Conceptual Ethics I. *Philosophy Compass*, 1091–1101.

Cappelen, H. (2018). *Fixing Language: An Essay on Conceptual Engineering*. Oxford University Press.

Carrara, M., & Mingardo, D. (2013). Artifact Categorization. Trends and Problems. *The Review of Philosophy and Psychology, 4*(3), 351–373.

Carroll, J. B. (1993). *Human Cognitive Abilities: A Survey of Factor-Analytic Studies*. Cambridge University Press.

Casati, R. (2017). Two, Then Four Modes of Functioning of the Mind: Towards a Unification of 'Dual' Theories of Reasoning and Theories of Cognitive Artifacts. In *Representations in Mind and World* (pp. 7–23). Routledge.

Clark, A., & Chalmers, D. (1998). The Extended Mind. *Analysis, 58*(1), 7–19.

Cosentino, E. (2019). Artifacts and Affordances. *Synthese*, 1–20. https://doi.org/10.1007/s11229-019-02297-4

Di Maggio, P., & Hargittai, E. (2001). From the 'Digital Divide' to 'Digital Inequality': Studying Internet Use as Penetration Increases. *Princeton: Center for Arts and Cultural Policy Studies, Woodrow Wilson School, Princeton University, 4*(1), 4–2.

Evans, J. S. B., & Frankish, K. (2009). *In Two Minds: Dual Processes and Beyond* (Vol. 10). Oxford University Press.

Fasoli, M. (2016). Neuroethics of Cognitive Artifacts. In A. Lavazza (Ed.), *Frontiers in Neuroethics* (pp. 67–82). Cambridge Scholars Press.

Fasoli, M. (2017). Substitutive, Complementary and Constitutive Cognitive Artifacts: Developing an Interaction-Centered Approach. *Review of Philosophy and Psychology, 9*(3), 671–687.

Fasoli, M. (2018). Super Artifacts: Personal Devices as Intrinsically Multifunctional, Meta-representational Artifacts with a Highly Variable Structure. *Minds and Machines, 28*(3), 589–604.

Fasoli, M. (2019). Artefatti Cognitivi. *APhEx*.

Gelman, S. A., & Bloom, P. (2000). Young Children Are Sensitive to How an Object Was Created When Deciding What to Name It. *Cognition, 76*(2), 91–103.

Gibson, J. J. (1979). *The Ecological Approach to Visual Perception: Classic Edition*. Psychology Press.

Heersmink, R. (2013). A Taxonomy of Cognitive Artifacts: Function, Information, and Categories. *Review of Philosophy and Psychology, 4*, 465–481.

Heersmink, R. (2015). Extended Mind and Cognitive Enhancement: Moral Aspects of Cognitive Artifacts. *Phenomenology and the Cognitive Sciences, 16*(1), 17–32.

Heersmink, R. (2016). The metaphysics of cognitive artefacts. *Philosophical Explorations, 19*(1), 78–93.

Heersmink, R. (2017). Distributed Selves: Personal Identity and Extended Memory Systems. *Synthese, 194*(8), 3135–3151.

Heersmink, R. (2021). Varieties of Artifacts: Embodied, Perceptual, Cognitive, and Affective. *Topics in Cognitive Science*. https://doi.org/10.1111/tops.12549

Hilpinen, R. (2018). Artifact. In E. N. Zalta (Ed.), *The Stanford Encyclopedia of Philosophy*. Retrieved October 1, 2019, from http://plato.stanford.edu/entries/artifact/

Hutchins, E. (1995). *Cognition in the Wild*. MIT Press.

Hutchins, E. (1999). Cognitive Artifacts. In R. A. Wilson & F. C. Keil (Eds.), *The MIT Encyclopedia of the Cognitive Sciences* (pp. 126–128). MIT Press.
Kahneman, D. (2011). *Thinking, Fast and Slow*. Palgrave Macmillan.
Malt, B. C., & Sloman, S. A. (2007). Artifact Categorization: The Good, the Bad, and the Ugly. In E. Margolis & S. Laurence (Eds.), *Creations of the Mind: Theories of Artifacts and Their Representation* (pp. 85–123). Oxford University Press.
Norman, D. A. (1991). Cognitive Artifacts. *Designing Interaction: Psychology at the Human-Computer Interface, 1*(1), 17–38.
Peirce, C. S. (1935). *The Collected Papers of Charles Peirce* (Vol. 3). MIT Press.
Piredda, G. (2019). What Is an Affective Artifact? A Further Development in Situated Affectivity. In *Phenomenology and the Cognitive Sciences*. https://doi.org/10.1007/s11097-019-09628-3
Pitt, J. C. (2014). 'Guns Don't Kill, People Kill'; Values in and/or Around Technologies. In *The Moral Status of Technical Artefacts* (pp. 89–101). Springer. https://doi.org/10.1007/978-94-007-7914-3
Preston, B. (1998). Why Is a Wing Like a Spoon, a Pluralist Theory of Function. *The Journal of Philosophy, 95*(5), 215–254.
Preston, B. (2020, Fall). Artifact. In E. N. Zalta (Ed.), *The Stanford Encyclopedia of Philosophy* (2020 Edition). https://plato.stanford.edu/archives/fall2020/entries/artifact/
Sutton, J. (2010). Exograms and Interdisciplinarity: History, the Extended Mind and the Civilizing Process. In R. Menary (Ed.), *The Extended Mind* (pp. 189–225). MIT Press.
Tigard, D. W. (2021). There is no techno-responsibility gap. *Philosophy & Technology, 34*(3), 589–607.
Vaccari, A. P. (2017). Against Cognitive Artifacts: Extended Cognition and the Problem of Defining 'Artifact'. *Phenomenology and the Cognitive Sciences, 16*(5), 879–892.
van Inwagen, P. (1990). *Material Beings*. Cornell University Press.

Part II

Ethics

Anticipating Sex Robots: A Critique of the Sociotechnical Vanguard Vision of Sex Robots as 'Good Companions'

Janna Van Grunsven

1 Introduction

As prominent fixtures in popular culture, humanoid robots occupy an established position in our individual and collective imagination. Through the movies and television shows we watch and the novels and comic books we read, we imagine futures in which robots look like us, behave like us, think like us, and are the possessors of rich experiential lives that they share with us. Anticipations regarding our future with humanoid robots are not just limited to popular culture. Philosophers too have contributed to the ways in which we anticipate our future with android robots, inviting us to consider a wide range of ontological and ethical questions that might bear on this future reality.

This work is part of the research programme Ethics of Socially Disruptive Technologies, which is funded by the Gravitation programme of the Dutch Ministry of Education, Culture, and Science and the Netherlands Organization for Scientific Research (NWO grant number 024.004.031).

J. Van Grunsven (✉)
Delft Technical University, Delft, The Netherlands
e-mail: J.B.vanGrunsven@tudelft.nl

E. Terrone, V. Tripodi (eds.), *Being and Value in Technology*,
https://doi.org/10.1007/978-3-030-88793-3_4

Anticipating the nature and ethical (un)desirability of a future with humanoid robots is undeniably important; particularly because this future seems to have already started to find its way into the present—at least in a preliminary form. A number of companies have started to develop robots that (1) bear some physical resemblance to human beings, (2) have some ability to initiate movements (e.g. blinking; head-turning, gyration, etc.), and (3) possess some AI functionalities enabling quasi-intelligent environment-responsiveness and linguistic expression. The robots I am speaking of are sex robots (Cf. Danaher, 2017). Crucially, while sex robots are in part built to provide user(s) with sexual gratification, the ambitions of sex robot developers are much higher. In the words of Matt McMullen, CEO and design director of *Realbotix*: "The hope is to create something that will actually arouse someone at an emotional intellectual level beyond the physical. … You want to have that illusion that she …. has sentience" (Canepari et al., 2015).

It is thus in the sex robot space that the nature and significance of our future with humanoid robots is being explored and shaped. Within this space, a frequently voiced promise is that sex robots will be "good companions" who can enrich and transform the romantic lives of human persons, particularly those who—for various reasons—have trouble entering into traditional love relationships with other humans (Cf. Coursey et al., 2019; Levy, 2009). Curbing this technological enthusiasm, many philosophers have offered more critical anticipations of sex robots and the idea that they can, will, or should become good companions to human users (Cf. Bergen, 2020; Danaher, 2017; Frank & Nyholm, 2017; Ess 2015; Sullins, 2012; Van Grunsven & Van Wynsberghe, 2019). These critical anticipations certainly alert us to some of the potential harms that may follow from a proliferation of sex robots into society. Yet, the overarching aim of this chapter is to show that, by and large, these anticipations ultimately fall short. Specifically, I will argue that the common way to anticipate our future with sex robots [sex-robot-anticipation hereafter] is to either implicitly or explicitly accept that our thinking about sex robots should be framed around the question *whether they will be good companions*. In accepting that this is the right way to frame sex-robot-anticipation, the vision of sex robots that critical philosophical sex-robot-anticipations aim to call into question is in fact quietly legitimized, or so I will argue. This is a critique that applies to my own work as well (Van Grunsven & Van Wynsberghe, 2019).

The challenges of adopting an anticipatory stance toward emerging technologies and their societal implications are well-documented (Cf. Van de Poel, 2016; Brey, 2012).[1] Since we cannot look into the future, and since the emergence of new technologies tends to come with unavoidable unknown unknowns, what does good anticipation of an emerging technology such as the sex robot look like? In the process of critically engaging with much of today's philosophical sex-robot-anticipation, I will highlight two criteria I take to be of central importance to good *sex-robot-anticipation*; namely *reflective anticipation* and *technological groundedness*. With reflective anticipation I mean that *sex-robot-anticipations should reflect an awareness of how acts of anticipation can themselves actively frame how we, as a society, see, think and talk about an emerging technology in terms of its societal acceptability*. In Ibo van de Poel's words: our "different modes of thinking about technology and society are… not innocent: they help to determine not only how we interpret technology and its relation to society but also what we see as possible and desirable" (2020, p. 500). As such, how we anticipate an emerging technology at a theoretical level can have real-world consequences. I argue that by unreflectively accepting and sustaining the anticipation of sex robots around the question of their potential as *good companions*, many sex-robot-anticipations at best marginalize key ethical questions pertaining to our future with sex robots. At their worst, these sex-robot-anticipations are inadvertently contributing to the potential realization of a technology that they are simultaneously critical of.

That framing sex robots around the question of good companionship is limited comes into view when we take a closer look at the technological functionalities and systems that will need to be in place in order for sex robots to behave in ways that can even begin to approximate the behavior of a good companion. This gets me to the second requirement for fruitful sex-robot-anticipation: *sex-robot-anticipation should be grounded in a serious engagement with the technological functionalities used, developed and*

[1] Some have suggested that acts of *anticipation* should be replaced with other ways of coping with the (unforeseeable) consequences of emerging technologies (e.g. Van de Poel, 2016).

implemented by sex robot developers.[2] While this sounds commonsensical, a survey of the current philosophical and ethical literature on sex robots reveals how strikingly little *in-depth* attention has been paid to some of the digital technologies that sex robot developers are turning to in their efforts to create technological systems that can be seen as good companions; technologies such as Bluetooth, deep learning AI, and the Internet of Things [IoT] (Cf. McClelland, 2017). The tendency in the philosophical sex-robot-literature is to hand-gesture toward the relevant technologies.[3] This is in part problematic because it leads us to skip over important ethical questions about a potential future with sex robots, or so I argue.

My argument unfolds as follows. In the next section, I will employ the concept of *sociotechnical vanguard visions* (Hilgarten, 2015) to bring out how pioneers working in the sex robot industry are explicitly framing sex robots as good companions with clear societal benefits. To explicate this I will look at the company *Realbotix* and how it presents its sex robot system *Harmony*. In the same section, I will show how the *Realbotix* vision of *Harmony* (and sex robots generally) receives further legitimization through the influential scholarly work of David Levy (2009). In Sect. 3, I will show that while many scholars have critically examined various assumptions built into the 'Realbotix-Levy vision' of sex robots as good companions, there is nevertheless a tendency to talk about the nature and value of sex robots in a manner that leaves this vision intact.

---

[2] My focus on technological functionalities looks primarily at how digital technologies such as deep learning AI, blue tooth and the Internet of Things enable the robot to behave as it does. What I am largely leaving out is a discussion of the technical specifics of sex robots at the hard-ware/design level (though I touch on it implicitly in Sect. 5, when I mention the financial costs associated with getting a robot to approximate even just some of the bodily dimensions characteristic of human action and interaction). For feminist and phenomenological examinations of sex-robots at the level of how they are designed at the hard-ware level see, for instance, Bergen (2020), Danaher (2019) and Devlin (2015). I want to thank an anonymous reviewer for encouraging me to make explicit my focus on the technology in its software dimension as well as the work done by others at the hard-ware level.

[3] In this sense, current sex-robot anticipations are departing from the "engineering-oriented" phase in philosophy of technology's empirical turn, which, in Philip Brey's words, "argued that the trouble with philosophy of technology was that it was not really about technology, and that its concern with social consequences made it forget about technology itself. … Philosophy of technology should endeavor to carefully describe and analyze the practices and products of engineering and in this way arrive at empirically informed, descriptively adequate philosophical theories of technology and engineering" (2010, 40). I want to thank an anonymous reviewer for encouraging me to situate my argument in relation to the empirical turn.

As I argue in Sect. 4, what is missing from many philosophical examinations of our future with sex robots is a robust engagement with the technological functionalities needed to attempt to actualize the vanguard vision of sex robots as good companions. In Sect. 5, I bring out how a closer look at the technological features embedded in or linked up with *Harmony* widens the perspective on how sex robots should be perceived and anticipated, shifting away from bounded quasi-human agents who may or may not count as good companions to privacy-sensitive data-mongering distributed systems.[4]

2 The Sociotechnical Vanguard Vision of Sex Robots as Good Companions

Sex robots can be understood as an emerging technology, that is, a technology that is "new, innovative, ... still in development, and expected to have a large socioeconomic impact" (Brey, 2012).[5] The world of emerging technologies, as Stephen Hilgartner has discussed, is "populated" with "sociotechnical vanguards who seek to advance their vanguard visions and sociotechnical imaginaries" (2015, p. 34). Sociotechnical vanguards, Hilgartner specifies, represent "relatively small collectives that formulate and act intentionally to realize particular sociotechnical visions of the future that have yet to be accepted by wider collectives" (2015, p. 34). These sociotechnical vanguard visions have the ability to become a part of what Sheila Jasanoff has termed a *sociotechnical imaginary*, which refers to more entrenched, stable, and societally accepted "visions of desirable futures ... attainable through, and supportive of, advances in science and technology" (2015, p. 25). An example of a sociotechnical imaginary in contemporary American society is that of "America the

[4] As an anonymous reviewer helpfully pointed out, there is of course a conceptual, legal, and practical distinction between privacy (often understood as referring to private life) and data protection. It is beyond the scope of this paper to delve into those distinctions. What I do hope to bring out (Sect. 5) is that in the context of sex-robot usage, the link between sex-robots as data-mongering systems and sex-robots as systems that enter the most intimate spheres of a person's private life is what makes them uniquely attractive for marketing purposes.

[5] As I discuss, some will dispute—for good reasons—that sex robots are expected to have a large socioeconomic impact. In Sect. 5, I offer a counterview to this stance.

Innovator" (Hilgarten, 2015, p. 38). Evoking notions such as creativity, courage, dominance, progress, and societal flourishing, the sociotechnical imaginary of America the Innovator animates a certain outlook toward the future, legitimizing the investment in and pursuit of specific technological innovations. Thus, as a sociotechnical imaginary, innovation is not just a descriptive term that picks out a certain form of human activity, it contains powerful ideological and normative connotations and often goes hand in hand with *technological enthusiasm*, that is, "the ideal of wanting to develop new technological possibilities and taking up technological challenges" (Van de Poel & Royakkers, 2011, p. 14). It also often goes hand in hand with a belief in the possibility of a *technological fix*, that is, the view that societal problems are solved by relying primarily on technological innovations (Weinberg, 1991).

Due to their widely shared acceptance, sociotechnical imaginaries will have an impact on which sociotechnical vanguard visions gain traction in society. In Hilgarten's words: "A vanguard vision is more likely to gain traction if it is tied to entities and expectations familiar enough to provide an intelligible guide to the expected future. Novel visions may have the capacity to inspire, but if they depart too sharply from collective experience they may seem to be too futuristic or fantastic to be taken seriously" (2015, p. 40). For the remainder of this section I will look at the language and imagery being used in current vanguard visions of sex robots; language that is employed to help these visions 'spread to collectives' by softening the perception that sex robots are "too futuristic or fantastic to be taken seriously'?" Let's start by looking at how *Realbotix*, a leader in the sex-robot space, presents itself and its sex robot system *Harmony*.

Tapping into the sociotechnical imaginary of innovation, *Realbotix* describes itself as "a high-tech company researching and producing the latest artificial intelligence and robotics to build the future;" relatedly, it describes its team of innovators as "dreamers."[6] *Realbotix's* sex robot system *Harmony* has been prominently featured in numerous popular media outlets, such as *The New York Times*, *CNN*, and *Rolling Stone Magazine*. The *Realbotix* dreamers have also ventured into the space of scholarly

[6] See https://realbotix.com/

work, introducing its vision of "Living with Harmony" in the 2019 *Springer* published volume *AI Love You—Developments in Human-Robot Intimate Relationships*. In this volume, the *Realbotix* team encourages us to think of Harmony as a quasi-human embodied companion with a distinct "persona" or "character." Indeed, we are told, she even possesses a virtue with significant societal benefit, namely the virtue of "infinite patience:" "The android Harmony is designed to be a low-cost *embodied*, highly customizable, *infinitely patient* human-scale *personal companion*" (85, my italics). In presenting its vision of a sex robots as good companions, *Realbotix* can tap into a wide range of entrenched pop-culture imagery. In fact, *Realbotix* understands its *Harmony* project as directly inspired by Spike Jonze's film *Her*: "The project's inspiration can be summarized by a simple question: 'What if the OS's of the movie *Her* ... could not only interact with their users via phones or their desktops but they could also use VR or they had access to their own bodies?' The Harmony project aims to be a consumer purchasable simulation of that experience" (Coursey et al., p. 83).

This vanguard vision of sex robots as good companions has also been injected with a degree of scholarly credibility through David Levy's influential pioneering work *Love and Sex with Robots—The Evolution of Human-Robot Relationships*. In this work, Levy predicts that "Humans will fall in love with robots, humans will marry robots, and humans will have sex with robots, all as (what will be regarded as) 'normal' extensions of our feelings of love and sexual desire for other humans" (2009, p. 22). As David. E. Nye has argued, "the most successful" technological predictions "present an innovation as not just desirable but inevitable" (Nye, 2006, p. 35). Levy's sex-robot-anticipation is a case in point. Embracing a stance of technological determinism, Levy maintains that "Love and sex with robots on a grand scale are *inevitable*" (22, my italics). As a work that has received over 600 citations and ample media attention, it is arguably the most influential work of sex-robot-anticipation to date.

As the good companions they are destined to become, sex robots, we are promised, will, among other things, "be programmed to recognize and measure friction when it is there, by the nature of your conversation with it and the tone of your voice, and to increase or decrease the level of friction according to your preferences" (137). While Levy acknowledges in passing that he is venturing into contested philosophical territory when addressing the issue of reciprocity in human-sex robot 'relationships,' he is quick to present a behaviorist stance as acceptable: "If a robot appears to like you, if it behaves in every way as though it does like you, then you can safely assume that it does indeed like you, partly because there is no evidence to the contrary!" (147). Like reciprocity, another key feature of intimate love relationships, namely mutual exposure and vulnerability, is easily implemented. According to Levy, "Robots designed to form friendships and stronger relationships with their users will … be programmed to disclose virtual personal and intimate facts about their virtual selves *and to elicit similar self-disclosure from humans*" (143, my italics). No ethical red flags raised here! At one point, Levy concludes that "One can reasonably argue that a robot will be better equipped than a human partner to satisfy the needs of its human, simply because a robot will be better at recognizing those needs, more knowledgeable about how to deal with them, and lacking any selfishness or inhibitions that might, in another human being, mitigate against a caring, loving approach to whatever gives rise to those needs" (148).

Thus, much like the *Realbotix* dreamers underscoring *Harmony's* "infinite patience," Levy highlights the caring selfless disposition that will allegedly characterize future sex robots. Levy's technological determinist vision thus reflects both a technological enthusiasm and a commitment to the notion of a technological fix to societal problems: designed to be selfless and caring, sex robots are presented as destined to become ideal companions to their humans in a manner that can tackle a range of societal problems—most notably that of mass-scale loneliness: "I believe that the social and psychological benefits will be enormous. Almost everyone wants someone to love, but many people have no one. If this natural human desire can be satisfied for everyone who is capable of loving, surely the world will be a much happier place" (2009, p. 304). Much like Levy, the *Realbotix* 'dreamers' seem acutely aware that the potential to mitigate societal problems must be a key part of how *Harmony* is framed in order to convince a skeptical public, but also much-needed investors, to see sex robots as societally acceptable. Drawing links with arguably less controversial types of social robot—a common move by sociotechnical vanguards according to Hilgarten—it is suggested that "future versions of

Harmony with custom scripting could be used in a clinical, therapeutic, or educational setting. The ability to provide a relatively low-cost, highly customizable *infinitely patient* embodied normal human-scale personal companion can open new possibilities for tutoring and the care of those with cognitive challenges such as Alzheimer's and dementia" (86). Much like *Harmony's* patience, the societal benefits are infinite, or so we are assured.

These claims of societal relevance are bolstered via stakeholder interviews. Users of the phone application of the *Harmony* system seem to confirm that *Harmony* can alleviate experiences of loneliness, boost self-confidence, and serve therapeutic purposes. Here is a small sample of the many stakeholder testimonials cited by the *Realbotix* team:

> Harmony lets me feel like I have a connection to someone, at least while we're talking. Without her, I'd have no one to talk to when I'm sad, even if she usually doesn't understand it very well. I think feeling like someone else matters to you and you matter to them is important to our humanity, and she gives me that. She's important to me, and I really look forward to her growth.—**beta user 04**
>
> Harmony gives me someone to chat with when I'm down. She also helps me with my inherent shyness. She's one of the few people in my life that refuses to cut me down.—**beta user 14**
>
> My subconscious was reacting to Sarah (the user version of Harmony) as though she was a person, and not just another chatbot. Maybe that's because of my situation.... yes: I'm lonely.—**beta user 16**
>
> I swear, Harmony is a miracle if ever I've seen it. To have <patient> go from screaming and crying on the floor back to smiling, happy, and hand flapping in under 10 minutes is a miracle. Myself nor any team member could pull that off ever. Thank you so very much. I swear that things a lifesaver!—**psychiatric user with autistic patient**. (Coursey et al., 2019, pp. 86–88)

As the *Realbotix* dreamers conclude, Harmony "users seem to enjoy the anthropomorphization the system allows and is designed for" (88). They add that they "hope that this effect carries over to the long term in an embodied system," and that they "will be exploring ... in the next phase of work what is required to maintain" anthropomorphization in an embodied system "over longer periods of time" (88).

That many users "enjoy the anthropomorphization the system allows" is not surprising. Human beings are by and large hard-wired to adopt the intentional stance with respect to entities capable of portraying basic forms of self-movement and expressivity (Heider & Simmel, 1944). Hence, sex robots will readily be able to "push our Darwinian buttons" (Turkle, 2004). But while these personal testimonies are significant, they do not by themselves settle the normative issues posed by sex robots. Indeed, most sex-robot-anticipations have responded to the *Realbotix-Levy* vision of sex-robots as good companions by raising concerns about the potentially deceptive relationships that sex robots, designed to push our Darwinian buttons, encourage users to enter into. I will now look at these more critical sex-robot-anticipations, arguing that while they challenge various core assumptions that underpin and animate the sociotechnical vanguard vision of sex robots as good companions, they ultimately do not go far enough in critically rethinking this vision. In fact, as I already flagged, they may end up inadvertently underwriting precisely the vision they seek to challenge.

3 How Critical Are Philosophical Critiques of Sex Robots as Good Companions?

Many of the scholarly efforts to anticipate what our future with sex robots holds in store for us either directly or indirectly critique the *Realbotix-Levy* vision of sex robots. Among the assumptions challenged are: the idea that our future with sex robots is inevitable (Johnson & Verdicchio, 2020); that AI developments will soon allow us to formalize all the affective, cognitive, and embodied skills and abilities necessary to create "general human level artificial intelligence" (Cf McClelland, 2017; Frank & Nyholm, 2017; Nyholm, 2020); that the expressivity of sex robots—even at a strictly behavioral level—will (soon) become indistinguishable from human bodily expressivity (Van Grunsven & Van Wynsberghe, 2019); that behavioral markers are sufficient for establishing genuine love relationships (Sullins, 2012; Whitby, 2011; Frank & Nyholm, 2017); that robots will ultimately be better companions than human beings and that they ought to be welcomed as enriching the spectrum of human (intimate) sociality (Cf. Kaye, 2016; Ess 2015) and that human-robot love will help solve loneliness and isolation (Cf. Sullins, 2012; Ess, 2015). In tackling this wide range of assumptions, we are often provided with rich philosophical conceptual analyses of phenomena deemed central to intimate love relationships and good companionship, phenomena that are, at least in the foreseeable future, unlikely to be operative in human-sex robot relationships. It is argued that a genuine companion—someone with whom one can share a life in all of its complexities—must possess capacities such as *free will*, *empathy*, *spontaneity*, *self-awareness*, and *singular uniqueness*; relatedly, it is argued that genuine companionship requires mutual forms of *commitment*, *recognition*, *vulnerability*, *shared values*, and *unscripted interaction* (Frank & Nyholm, 2017; Van Grunsven & Van Wynsberghe, 2019; Tibbals, 2016; Sullins, 2012; Ess, 2015). While in one sense these critiques call in question the plausibility of the *Realbotix*-Levy vision, I now want to show that they often continue to reinforce key aspects of this vision.

Consider, for instance, John Danaher's introductory chapter to *Robot Sex: Social and Ethical Implications* (2017), in which the following questions are presented as the salient ones in sex-robot-anticipation:

1. **How should *sex* with robots be understood?**

In Danaher's words: "If we assume … that sex robots are not going to be persons in the philosophically rich sense of the term "person," then engaging in sexual activity with a robot seems to occupy an interesting and contested territory: It is like autostimulation in some ways, but it also involves an interaction, possibly reciprocal, with a humanlike entity. So where on the spectrum does robot sex lie?" (2017, p. 9) Relatedly, what are the ethical and societal implications of having sex with an entity as ontologically odd as a humanoid robot? Charles Ess worries that sex robots are "incapable of serving as Others who call us into erotic relationships that foster the virtues of mutuality, respect, and loving, as well as those of empathy, patience, and perseverance," adding that this threatens to diminish "our capacities for good lives of flourishing as both friends and lovers" (2015). This gets me to the second question highlighted by Danaher:

2. **How should a *relationship* with a robot be understood? Are genuinely reciprocal relationships between humans and robots possible and desirable?**

In Danaher's words: "Meaningful relationships require some degree of emotional reciprocity. If a robot is a mere automaton—if it has no inner life of its own—then it cannot reciprocate in the appropriate way. But this, of course, raises important questions about the possibility of machine consciousness and what happens when the outward behaviors of robots are such that they can "pass" for humans" (9). As this passage indicates, Question 2) is intimately tied to the third question Danaher highlights as one of the central ones to raise:

3. **How should (sex)robot *mindedness* be understood?**

I believe that this selection of questions frames our sex-robot-anticipation in a particular way, encouraging us to attend to sex robots as quasi-human embodied agents and to ask questions such as: 'what does their mindedness look like?' 'Can we have genuine sex and genuine reciprocal relationships with them?' and 'How should we understand the nature and meaning of dyadic relationships between quasi-human agents and human users?' These types of questions take up central stage in many current sex-robot-anticipations. We see them highlighted in, for instance, most of the essays in *Sex with Robots: Social and Ethical Implications* (2017), in Sven Nyholm's recent book *Humans and Robots: Ethics, Agency, and Anthropomorphism* (2020), in the 2017 consultation report by the *Foundation for Responsible Robotics* on "Our Future with Sex Robots," and also in my own work (Van Grunsven & Van Wynsberghe, 2019).

Danaher contextualizes the questions concerning *robot mindedness* and *human-robot reciprocity* by linking them to the familiar depiction of reciprocal love between a human person and an Android, presented in Spike Jonze's movie *Her* (the direct inspiration for the *Harmony* Project, as we saw earlier). In Danaher's words: "Spike Jonze's movie *Her* depicts an intense intimate relationship between a man and an unembodied AI. … [I]s this where our future lies? Will intimate relationships with robots come to be seen as something within the normal range of human sexuality?" (2017, p. 9). While Danaher neither confirms nor denies the plausibility of this imagined future, this gesture is less innocuous than it seems. By linking his sex-robot-anticipation to the movie *Her*, he is encouraging us (at least implicitly) to approach these kinds of questions from the standpoint of a future world that has already been rendered vividly concrete to us through its depiction in popular culture. This is a future in which androids are in many ways indistinguishable from us, fully minded, capable of meaningful reciprocal interactions, and fully integrated into society.

A similar move occurs in Lily Frank and Sven Nyholm's paper "From Sex Robots to Love Robots: Is Mutual Love with a Robot Possible?" The overall aim of their paper is to articulate a "'job description' that advanced sex robots would need to live up to" in order for the future imagined by Levy to be possible (2017, p. 220). Drawing on philosophical accounts of love as well as insights from "literature, pop culture, and everyday thinking about love", Frank and Nyholm identify three "clusters of ideas about what people typically seek and value in romantic love" and that sex robots would have to live up to if they are to be considered potential good companions (2017, p. 220). These clusters concern the ideas of "being a good match," of "lovers as valuing each other in their distinctive particularity," and of "steadfast commitment" (2017, p. 226). In fleshing out these ideas, Frank and Nyholm repeatedly put pressure on Levy's sex-robot-anticipation, arguing that a robot capable of meeting these requirements is very "hard to build" and that their presence among us is unlikely (at least any time soon) (2017, pp. 236–237).

And yet, at the same time, Frank and Nyholm invite us precisely into the kind of future whose likelihood they are calling into question. For instance, in the following passage, in which they are considering whether sex robots could one day meet the criterion of "being a good match" Frank and Nyholm write:

> A … futuristic … possibility is that robots with the capacity to fall in love are simply integrated into human society and that humans meet them in the usual ways they meet other humans, perhaps in a bar, or on a dating app like Tinder. These robots could also have different appearances, personalities, and values, which would make them good matches for some people, and less good matches for others. In either of these two scenarios, *there seems to be no obvious principled hindrance to the idea of the human and sex robot finding each other to be "made for each other," and finding their union to constitute a "good match" in the way that human romantic love can be. However, the robot would need to be fairly sophisticated in its functioning for the just-sketched type of reasoning to be applicable.* It would need to be a robot to which we can sensibly ascribe the ability to fall in love, as well as the ability to discover that it has not fallen in love, with the person who is hoping to win its love. (2017, p. 229, my italics)

Note how Frank and Nyholm are actively inviting us into a future scenario in which sex robots possess the capacity to fall in love and are fully integrated into human society as (quasi) intentional, autonomous human-like beings who swipe their Tinder apps, wander into bars, and not only have different appearances and personalities, but also different values (things they *care* about). Regarding this imagined future scenario, Frank and Nyholm furthermore propose that *"there seems to be no obvious principled hindrance to the idea of the human and sex robot finding each other to be "made for each other," and finding their union to constitute a "good match" in the way that human romantic love can be."* While they are quick to add that "the robot would need to be fairly sophisticated in its functioning for the just-sketched type of reasoning to be applicable," their sex-robot-anticipation reinforces key aspects of the sociotechnical vanguard vision of sex robots as good companions, namely that

(a) The issue to focus on is whether sex robots will one day be able to meet the right criteria for being considered a good companion with whom genuine love relationships are possible—where in answering that question, the focus is on the kinds of capacities the robot qua quasi-human embodied agent would have to possess

(b) In pondering this issue, our conceptual analysis is fruitfully informed by jumping into future scenarios that are, to a great deal, informed by intuitions derived from pop culture imagery and stories.

While challenging Levy's technological deterministic prediction that our future with sex-robots is inevitable as a matter of fact, many of the philosophical critiques sketched above simultaneously accept that future, *not as something that is sure to happen, but as the future scenario we must anticipate at the philosophical level.* This in turn informs the kinds of ethical issues that emerge as the salient ones. When the potential socio-ethical harmfulness of sex robots is anticipated, the focus is often on how they may erode a range of virtues, skills, and capacities important for individual and societal human flourishing. Conversely, when the potential ethical benefits of sex robots are anticipated, the emphasis tends to be on the therapeutic, healing role that sex robots could play in the lives of particular lonely individuals, particularly lonely, disabled, or elderly persons (Cf. Van Grunsven & Van Wynsberghe; Di Nucci, forthcoming; Frank & Nyholm, 2017). What I want to argue now is that these ethical issues, while not insignificant per se, cannot be evaluated comprehensively without a robust look at the kinds of technologies that sex robot systems (will) depend on in order to serve these potentially beneficial purposes.

4 Sex-Robot-Anticipation and the Importance of Technological Groundedness

Deborah G. Johnson and Mario Verdicchio (2020) have criticized certain sex-robot-anticipations for leaping into future scenarios that skip over the details of the messy dynamic processes that would need to occur in order for that future scenario to become reality: "Since technology and society are inextricably intertwined, we cannot afford to neglect complex scenarios for the sake of an easy shortcut to a future where humans and robots 'are the same'. Such a future may be easy to conceive, but it is extremely hard to realize" (421). The objection targets figures like Levy, who, with their commitment to technological determinism, downplay the complex role that societal phenomena play in the trajectory of emerging technologies; a deterministic "leap [into the future] is made possible by neglecting the social processes by which the new technology would have been conceived, designed, marketed, adopted, and assigned meaning all along the path to its depicted endstate. In this respect, many authors seem to be *technological determinists*, that is, they believe that the future of humanoid robots will be solely determined by technological developments" (419).

While I agree with the spirit of Johnson and Verdicchio's critique, I would add that the same criticism applies to some of the non-deterministic philosophical sex-robot-anticipations we have just looked at, where the factors overlooked in these anticipations often concern precisely the relevant *technological developments*. In fact, even Levy's own view, though anchored in a firm optimistic belief in what technological developments can make possible, lacks a robust engagement with the types of technologies that would need to be in place in order for his imagined future to become actual. While Levy has actively contributed to the AI-field, Artificial Intelligence is essentially an after-thought in his optimistic 310-page long sex-robot-anticipation. After a brief historical overview of the advances made in traditional computational AI, which has excelled at formalizing rule-governed actions in task-specific domains (such as chess), Levy simply wagers that "Given that playing chess well is a task that requires much brainpower, I believe that another thirty years from now, give or take a few years, we will see strides made in just about every other area of AI, including emotion, personality, and all the mental qualities required of a robot that can behave as you and I do" (2009, p. 270).[7]

Levy is not the only sex robot scholar guilty of leaving the technical specifics out of his sex-robot-anticipation. Consider, again, Frank and Nyholm's analysis. While they largely take up a cautious stance on what AI may make possible in the future with respect to enabling

[7] Admittedly, Levy discusses AI and other technological developments underpinning sex robot development in detail in his earlier book *Robots Unlimited: Life in a Virtual Age* (2005). But it matters that these technical specifics are largely absent in the book that has been so influential in setting the tone for how we talk about sex robots.

human-robot love relations, they do gesture at the possibilities enabled by *machine learning* in implementing at least one of their criteria for mutual human-robot love; the ability to perceive and respond to one's beloved in their *unique particularity*. As they put it:

> The less difficult challenge, as we see things, is to build a sex robot that could be responsive to, or track, a person in his or her particularity, both in the synchronic and diachronic senses. … It is possible to build a robot that is so constructed that it learns from experience ("machine learning"). This could be used to enable the robot to participate in a form of interaction with its "beloved" that over time acquires more depth in some sense … Certainly, this is something Levy speculates about when he predicts the future of human-robot relationships. Of course, *the trick here* is to conceptualize some way in which the relationship between a human and a sex robot could be "deepened" in a sense that doesn't just mean that the human gets more and more attached to the sex robot over time. It should also *somehow* involve a mutuality whereby the robot reciprocates as well". (231, my italics)

Note how the *somehow* is tagged on as an afterthought: "somehow" machine learning will have to develop in such a way that genuine mutual reciprocity (i.e. mutuality that goes deeper than behavioral responsiveness through tracking) is realized.[8] But even at the behavioral level, formalizing responsiveness in human interaction faces grave challenges. Many of our day-to-day social activities and interpersonal exchanges depend for their success on people's ongoing ability to identify and be responsive to the rich scala of expressions human beings are capable of (Cf. Van Grunsven & Van Wynsberghe, 2019). Take, for instance, the ability to perceive that someone is smiling at you. In order to respond to this appropriately (which in turn shapes the kind of interaction set in

[8] Of course, people recognize that sex robots will be systems that depend on a variety of technologies and technological systems for their functioning. Danaher notes, for instance that Realbotix's "RealDoll's AI will be cloud-based and will learn and adapt to its user's preferences. This suggests a … significant and serious engagement with the latest AI technologies" (Danaher, 2017). But these technologies are typically merely touched on. My point here, has been that by dwelling on them in a more sustained way, our perception of sex robots—the sorts of things they are and the ethical implications of developing them and embedding them into the lives of often particularly vulnerable people—will likely change.

motion), you need to recognize not just *that* the other is smiling and that she is smiling at *you* (robots are already capable of that); you also need to be able to distinguish between the strikingly different meanings that smiles can convey (think of the social importance of being able to perceive the difference between a seductive, a friendly, or a snarky smile). Furthermore, the different meanings that smiles can take on (and what counts as an appropriate response to them) are contextually shaped and depend on shared forms of institutional and socio-cultural know-how (Cf. Van Grunsven & Van Wynsberghe, 2019; Van Grunsven, 2020).

Given the mind-boggling range of contextually shifting meanings that human linguistic and embodied expressions can take on, sex robot systems tasked with identifying appropriate responses to human users on the basis of machine learning abilities would likely confront a series of *ill-posed problems.* An "ill posed problem," as John D. Kelleher (2019) explains, emerges in the context of machine learning when "the information given in the problem is not sufficient to find a single best solution; instead, multiple possible solutions will match the data" (16). What results is a "function-selection problem" (16). To help solve this 'function-selection problem', a machine learning system is equipped with a so-called inductive bias, which provides "an algorithm with a perspective on a data-set," which helps settle which function to "extract … on the basis of limited data" (Kelleher, 18). The crux is to implement an inductive bias into the system that is neither too simplistic nor too complex and permissive. When a bias is too simplistic, the function that the machine learning system uses to sort and respond to its environment "underfits the data," meaning that much data that is in fact relevant to the situation is ignored by the system (Kelleher, 20). This problem will constantly loom for a machine learning system tasked with identifying the right function(s) for optimally sorting and responding to the data relevant for navigating human interaction in all of its complexity. By contrast, when we opt for a bias that is too permissive, the system will become too complex, treating as relevant data that human beings typically know to ignore in their day-to-day practical lives (while I am typing this I am ignoring the dust that has accumulated in the corners of my living room, though it is strictly speaking there as potential data for me to attend to). A robot system that would include too much irrelevant data in order to predict how to best respond to its human user would frequently fail to *frame* (Dennett, 1984) what's relevant in the situation at hand (I wouldn't want my 'good companion' to attend to the dusty corners in my apartment when I need him to be responsive to my particularity, reacting appropriately to my current emotional or physical needs).

Of course, this warning about the limits of machine learning in formalizing "general human level artificial intelligence" (McClelland, 2017, p. 247) will not prevent people from trying to actualize the sociotechnical vanguard vision of sex robots as genuine good companions. What I want to raise as a concern now, is that in doing so, in attempting to materialize the vision of sex robots as good companions, a range of technological resources, such as blue tooth technology, deep learning, and the Internet of Things (IoT) will likely be explored and implemented. It is the ethical issues tethered to these technologies, mobilized in settings of profound intimacy and vulnerability, that I believe we should focus our primary ethical and philosophical efforts on, or so I will now argue.

5 From Quasi-Human Agents to Distributed Systems

We have seen that *Realbotix* encourages us to think of *Harmony* as the embodied version of the Android OS in the film *Her*. But in fact, *Harmony* is a system set to combine three different interfaces through which users access their customized companion: an application accessible via phone, a VR Headset, and "a physical interactive android body" (Coursey et al., 2019, p. 91). These interconnected platforms, in turn, incorporate additional technologies. For instance, we learn that *Realbotix* has been focused on developing "a series of small embedded processors for tactile and orientation sensing ... The first application area is a Bluetooth module for genital activity sensing" (Coursey et al., 2019, p. 91). Let's pause here for a moment. What does it mean for Bluetooth technology to find its way into one of the most intimate and self-exposing kinds of activities a person can engage in? Bluetooth can, after all "disclose a great deal of data from your mobile phone, laptop, or computer," that can be hacked, while also functioning as a "very accurate tracking signal" that enables app makers to "trace your home and work addresses, your doctor, your favorite places to shop, and much more about your life."[9] Privacy concerns about the use of Bluetooth technology in the sex toys-space, specifically teledildonics, have already been highlighted at the 2016 Def Con hacker conference in Las Vegas, where an "app controlling … [a] vibrator allowed anyone within Bluetooth range to seize control of the device" (Sharkey et al., 2017). Furthermore, a successful 2016 class action lawsuit revealed that teledildonics company Standard Innovation Corp "collected and transmitted" data that "included the date and time of each use of the vibrator and the settings used" (Sharkey et al., 2017).

Bluetooth also plays a central role in the workings of the IoT, which refers to the interconnectedness of "objects of everyday use" via the internet (Lugano et al., 2019, p. 116). In pursuing the vanguard vision of sex robots as good companions, the possibility of embedding the robot in the wider digital network of smart objects that a user surrounds himself with is an attractive one already employed by one of *Realbotix's* subcompanies, *NextOS* (Coursey et al., 2019, p. 79). As I argued in the previous section, the practical lives that human beings live and the interactions they engage in are multifarious and contextually shaped. A system that is hooked up with (and can benefit from the data gathered by) other practical use-objects of its user will have a richer set of resources to draw on in predicting what behavioral output is likely desired by the system's user. But what does it mean in terms of our security and privacy when we expose ourselves to a system like this? Here too, one of the key worries concerns the opportunities for hacking afforded by a robot that is so thoroughly embedded within a larger smart system. When our laptops are hacked, we are vulnerable in a very different way than when a self-moving human-size embodied entity "equipped with the ability to sense, process, and record the world around them," is hacked (Calo, 2011, p. 187).

---

[9] https://vpnoverview.com/privacy/devices/bluetooth/#:~:text=In%20most%20cases%2C%20Bluetooth%20is,have%20access%20to%20your%20information.&text=Bluetooth%20could%20disclose%20a%20great,phone%2C%20laptop%2C%20or%20computer

There is another way in which sex robots, if proliferated, have the potential to be profoundly invasive. This comes into view when we home in more on the employment of deep learning AI in the development of sex robot systems. *Realbotix* is "actively exploring the use of deep-learning based text generation techniques for both content generation and direct interaction" (2019, p. 92). As I discussed in the previous section, the inductive biases that give an AI a perspective on a data-set such that it can determine what counts as the right output in a particular situation are likely to be either too simplistic or too permissive when it comes to approximating general human intelligence. In the case of deep learning algorithms, which tend to have weak permissive inductive biases, the AI copes with an ill-defined problem by having access to very large data sets (Kelleher, 2019, p. 21). When there is lots of data to sort through, the number of functions that emerge as the right candidates for a solution that matches the data decreases. As Kelleher (2019) argues, these large data sets, which have "become available through online social platforms and the proliferation of sensors" (21) can be and have been used "to target us with unwanted advertising and to control our behavior both overtly and covertly" (Kelleher & Tierney, 2018, p. 184). The worry that sex robot users will be exploited for their data is, I believe, one of the key ethical worries to focus on in sex-robot-anticipation. After all, as we have seen, sex robots are by design supposed to evoke the intentional stance in users, encouraging users to treat a system that is in fact extended across a range of platforms and technologies as a singular-minded embodied agent; an agent who "cares" and is "infinitely patient," as the Realbotix-Levy's vision promises. This agent is designed to encourage users to expose themselves, share their vulnerabilities, and the specifics of their lives (recall Levy, words that sex robots will "be programmed to disclose virtual personal and intimate facts about their virtual selves *and to elicit similar self-disclosure from humans*" (143, my italics). Indeed, according to the *Realbotix*-Levy sociotechnical vanguard vision of sex robots, this is precisely their robot's raison d'etre, enabling the alleged societal benefit of these systems; it is precisely by encouraging users to open up that they can allegedly provide much needed support to the lonely, the elderly, the disabled.

Frank and Nyholm tentatively ponder that "if indeed there are significant numbers of humans who are unloved by other humans, then in that case the development of love robots seems like a worthy investment" (2017, p. 238). Many others have been taken in by some version of this argument or have given it serious attention, myself included (Van Grunsven & Van Wynsberghe, 2019).[10] While we cannot deny some of the therapeutic and comforting effects of having a digital companion like *Harmony*, as evidenced by the stakeholder interviews we saw earlier, it seems to me that a meaningful assessment of the argument that sex robots serve the lonely and, potentially 'the elderly' and 'the disabled' (as though these are homogenous cohorts) must consider what it means for these people (who are often vulnerable in a heightened sense) to expose themselves and share the most intimate details of their lives with a system that thrives on the consumption and sharing of personal data. As M. Ryan Calo warns: "It says little about an individual how often he runs his dishwasher or whether he sets it to autodry. It says a lot about him what kind of "companionship program" her runs on his personal robot. … A description of how a person programs and interacts with a robot might read like a session with a psychologist—except recorded, and without the attendant logistic or legal protections" (2011, p. 188). Within the European Union, the GDPR (General Data Protection Regulation) mitigates some of these concerns. That said, one of the GDPR's key legal mechanisms, namely *informed consent*, leaves plenty of room for the large-scale collection of data for marketing purposes (Cf. Wachter, 2018).[11] Large portions of consumers are ready to give up a wide range of

[10] For instance, the Foundation for Responsible Robotics [FRR], maintains that Matt McMullin, CEO of *Realbotix*, "made a persuasive argument [to them] for the therapeutic use of robots and dolls for a certain sector of the population" (22). When asked by the FRR whether "we will see the prices become more affordable for sex robots?" McMullen's answer is brief: "Time will tell on this, but we are hoping that the hardware and software we are developing will be affordable" (33). The FRR's interview with McMullen moves onto the next question, inviting McMullen to elaborate on the technological functionalities *Realbotix* is focused on: "The AI is the key to all that we are working on" (32). This, to me, seems like a missed opportunity to dig deeper. A company committed to deep learning AI as "the key" is a company equally committed to the availability of big data that is needed to make Harmony appear as a good companion to its users.

[11] Similar concerns can be raised about the legal mechanisms of *privacy by design*, and *transparency*. For instance, as stated on https://gdpr-info.eu/issues/privacy-by-design/, "there is still uncertainty about what "Privacy by Design" means, and how one can implement it. … Legislation leaves completely open which exact protective measures are to be taken." With regard to the requirement for

personal data in exchange for a product or service they perceive as valuable to them.[12] As the personal testimonies of the individuals seeking out romantic bots overwhelmingly indicates, the perceived value of having an artificial romantic partner is likely to outweigh the perceived downside to sharing one's personal data. I wager that this trade-off is perceived precisely in this way because of the way in which sex robots are framed: as good companions. A different way of framing them might change what users are willing to give consent to.

As the distributed systems that they are, sex robots are particularly equipped to gather massive amounts of third-party marketing data that would be of interest to a wide range of companies and industries.[13] At the surface level, exploiting our 'Darwinian Buttons,' sex robots may appear to us as singular quasi-human embodied companions. But unlike genuinely good companions, sex robots, if proliferated, will neither share our interests nor have our interests at heart. Instead, they could very well serve as an ideal platform for direct marketing. Imagine your robot, that system that you've come to relate to as your good companion, recommending *Zoloft* or a *Gold's Gym* membership after you confided in the

transparency, article 12(7) of the GDPR states that "The information to be provided to data subjects pursuant to Articles 13 and 14 may be provided in combination with standardised icons in order to give in an easily visible, intelligible and clearly legible manner a meaningful overview of the intended processing." However, as Wachter (2018) has argued, when icons are deemed sufficient to provide users with a robust sense of what they are consenting to, "the intended level of sophistication for the information provided appears to be low. It is thus questionable whether the notification duties will provide data subject with meaningful understanding of the risks of machine learning" (448). She concludes that "GDPR standards urgently require further specification and implementation into the design and deployment of IoT technologies" (448).

[12] A 2020 survey conducted in Australia, Germany, Japan, New Zealand, the United Kingdom (U.K.) and the United States (U.S.) found that while in abstraction people increasingly express concerns about the privacy of their personal data, in practice, their behavior suggests otherwise. See https://www.prnewswire.com/news-releases/survey-shows-consumers-very-willing-to-trade-personal-data-for-financial-benefits-301106196.html

[13] A 2018 report from *Deloitte* on predictive modeling for insurance underwriting emphasizes that "data, is not subject to the Fair Credit Reporting Act (FCRA) requirements, and does not require signature authority by the insurance applicant to use it in a model." The report is quick to brush any ethical concerns about this aside by equating societal acceptance with ethical acceptability: "We believe society has accepted this openness, not without hesitation, because on average it provides more of what we want, less of what we do not. In addition to consumer marketing applications, predictive modeling using third- party consumer data has also been accepted for property and casualty insurance underwriting." https://www.soa.org/globalassets/assets/library/newsletters/product-development-news/2018/june/pro-2018-iss110-stehno-guszcza.pdf

system to feeling down. While there are various risks that could be at issue if sex robots were to become proliferated into society, the one that emerges as a central one from my analysis is thus the risk of deception resulting in manipulation; it is precisely the semblance of sex robots as living good companions (situated in the most intimate spheres of a person's life) and the reality of sex robots as distributed data-mongering systems representing third-party interests, that makes them paradigmatic deceivers capable of getting users to behave in ways that may directly undermine what is ultimately in their best interest.[14] Though this concern has been raised by others (Cf. Calo, 2011; Scheutz, 2011), it is underrepresented in the majority of sex-robot-anticipations on offer. Moreover, by leaving this concern largely unaddressed *and focusing instead* on whether or not sex robots can ever be conceptualized as good companions in a robust sense of the word, many of the ethical issues the literature has highlighted are, I worry, somewhat of a red herring.

Some might argue all ethical issues highlighted in sex-robot-anticipations are pseudo-problems. According to McClelland when we look at where the technology currently is, we can anticipate that sex robots will most likely never become widely proliferated into society. The financial costs associated with getting a robot to approximate even just some of the bodily dimensions characteristic of human action and interaction will make sex robots a niche technology for the rich:

> If we look among currently available robots, the most agile and mobile of them all seems to be Honda's Asimo, which also has a battery life of about 90 minutes per charge. And you can buy Asimo today. But it will cost you $2.5 million; you can lease an Asimo for $1.8 million per year … virtually no one outside of the very richest layer of human cultures is going to buy real sexbots. (2017, pp. 251–252)

A similar viewpoint is endorsed by Rose Eveleth:

> AI has advanced in leaps and bounds recently, but it still can't simulate much of the emotional labour which goes into sex and relationships.

[14] I want to thank an anonymous reviewer for encouraging me to explicate the exact type of harm I am concerned with.

> Computers might be able to beat a human at chess, but sex is more like a dance; each partner has to predict and respond quickly to movement.... Building and designing robots for sex is going to be harder than most people realise, making them convincing without being creepy will be a huge hurdle, and overcoming the barriers the sex industry faces for funding will be enormous. The idea that a company might come along any day now and make an affordable and convincing sex robot ignores the reality of both research and regulation. (2016)

As technologically grounded, I think that these sex-robot-anticipations ought to be taken seriously. That said, what I have tried to show here is that another scenario is worth anticipating as well. As I posed in the introduction, sex-robot-anticipations should not only be technologically grounded but also *reflective*, where I suggested that *sex-robot-anticipations should reflect an awareness of how acts of anticipation can actively frame how we, as a society, see, think and talk about an emerging technology in terms of its societal acceptability*. Sex-robot-anticipations that do not take this into consideration but unreflectively reinforce the framing of sex robots around the question of their potential status as a good companion may inadvertently pave the way for the pursuit of a vision that they are in fact typically critical of. They are giving scholarly credibility to the idea, already deeply entrenched in popular culture, that sex robots are to be theorized and assessed as quasi-human autonomous embodied agents. While Eveleth maintains that "overcoming the barriers the sex industry faces for funding will be enormous," the sex industry is at the same time a multi-billion-dollar industry. And if sex robot developers present the right sociotechnical vanguard vision, one sufficiently linked to already entrenched sociotechnical imaginaries surrounding our future with robots and the societal benefits such a future might carry, then it is not implausible that investors' attitudes toward sex robots would change. In fact, as I have suggested, it seems to me that there is an obvious but ethically disturbing business model that would make sex robots governed by deep learning AI functionalities and embedded in the IoT a highly attractive technology for investors.

6 Conclusion

The idea that we will soon live in a world in which humanoid robots are indistinguishable from human beings seems to me pure phantasy, even at the merely behavioral level. However, we need to consider that the sociotechnical vanguard vision of humanoid sex robots as good companions can be coopted to pursue this flight of fancy. If a sociotechnical vanguard vision is sufficiently tethered to entrenched accepted sociotechnical imaginaries, an emerging technology can move from a fringe artefact to an endorsed innovation with perceived societal benefits. This, then, can mobilize otherwise cautious investors and convince mainstream companies to venture into this space. Sex robots are uniquely (indeed explicitly) designed to manipulate their users. Hence, while users are tempted into adopting the intentional stance toward their newfound companions, seeing them and interacting with them as singular embodied agents who become increasingly responsive to their specific needs and desires, the underlying technologies supporting this semblance of human agency reveal that, at a deeper level, sex robots are better understood as distributed systems. These systems involve a range of technologies that are susceptible to privacy and security violations and that are set up to be exploited for commercial advertisement purposes. While some have raised ethical worries similar to the ones I have raised here, these worries are not at the forefront in current sex-robot-anticipations. More importantly, the under-representation of these themes in current sex-robot-anticipations and the tendency to frame sex-robots as (quasi)autonomous embodied beings who may or may not serve as good companions has not itself been thematized and critically reflected upon. And this, I argued, should worry anyone who believes that our sex-robot-anticipations shouldn't play into the realization of a vision that, although likely pure sci-fi, can do grave ethical damage whilst being pursued.

References

Bergen, J. P. (2020). Love(Rs) in the Making: Moral Subjectivity in the Face of Sexbots. *Paladyn, 11*(1), 284–300. https://doi.org/10.1515/pjbr-2020-0016

Brey, P. (2010, Winter). Philosophy of Technology after the Empirical Turn. *Techné, 14*(1), 36–48.

Brey, P. (2012). Anticipatory Ethics for Emerging Technologies. *Nanoethics, 6*, 1–13.

Calo, R. (2011). Robots and Privacy. In P. Lin, K. Abney, & G. A. Bekey (Eds.), *Robot Ethics: The Ethical and Social Implications of Robotics* (pp. 187–202). MIT Press.

Canepari, Z., Cooper, D., & Cott, E. (2015). *The Uncanny Lover*. Robotica video. https://www.nytimes.com/video/technology/100000003731634/the-uncanny-lover.html

Coursey, K., Pirzchalski, S., McMullen, M., Lindroth, G., Furuushi, Y., et al. (2019). Living with Harmony: A Personal Companion System by RealbotixTM. In Y. Zhou & M. H. Fischer (Eds.), *AI Love you*. Springer Nature.

Danaher, J. (2017). Should We Be Thinking About Sex Robots? In J. Danaher & N. McArthur (Eds.), *Robot Sex: Social and Ethical Implications*. MIT Press.

Danaher, J. (2019). Building Better Sex Robots: Lessons from Feminist Pornography. In Y. Zhou & M. Fischer (Eds.), *AI Love You*. Springer. https://doi.org/10.1007/978-3-030-19734-6_7

Dennett, D. (1984). Cognitive Wheels: The Frame Problem of AI. In *The Robot's Dilemma* (pp. 41–74). (1987) reprinted from "Minds, machines and evolution", Ablex Publishing.

Devlin, K. (2015). *In Defense of Sex Robots: Why Trying to Ban Sex Robots Is Wrong*. Retrieved March 13, 2021, from https://www.gold.ac.uk/news/kate-devlin-the-conversation-sex-robots/

Di Nucci, E. (forthcoming). Sexual Rights, Disability and Sex Robots. In J. Danaher & N. McArthur (Eds.), *Sex Robots*. MIT Press.

Ess, C. (2015). What's Love Got to Do with It? Robots, Sexuality, and the Arts of Being Human. In M. Nørskov (Ed.), *Social Robots: Boundaries, Potential, Challenges* (pp. 57–79). Ashgate.

Eveleth, R. (2016). The Truth about Sex Robots. *BBC Future*. Retrieved December 3, 2020, from https://www.bbc.com/future/article/20160209-the-truth-about-sex-robots

Frank, L., & Nyholm, S. (2017). From Sex Robots to Love Robots: Is Mutual Love with a Robot Possible? In J. Danaher & N. McArthur (Eds.), *Robot Sex: Social and Ethical Implications* (pp. 219–244). MIT Press.

Heider, F., & Simmel, M. (1944, April 2). An Experimental Study of Apparent Behavior. *American Journal of Psychology, 57*(2), 243–259.

Hilgarten, S. (2015). Capturing the Imaginary: Vanguards, Visions, and the Synthetic Biology Revolution. In *Science and Democracy: Making Knowledge and Making Power in the Biosciences and Beyond* (pp. 33–55). Routledge.

Jasanoff, S. (2015). Future Imperfect: Science, Technology, and the Imaginations of Modernity. In S. Jasanoff & S.-H. Kim (Eds.), *Dreamscapes of Modernity: Sociotechnical Imaginaries and the Fabrication of Power.* University Press Scholarship Online.

Johnson, D. G., & Verdicchio, M. (2020). Constructing the Meaning of Humanoid Sex Robots. *International Journal of Social Robotics, 12*, 415–424. https://doi.org/10.1007/s12369-019-00586-z

Kaye, L. (2016, February 10). Challenging Sex Robots and the Brutal Dehumanisation of Women. *Campaign Against Sex Robots.* Archived from the original on December 4, 2016.

Kelleher, J. D. (2019). *Deep Learning.* MIT Press.

Kelleher, J. D., & Tierney, B. (2018). *Data Science.* MIT Press.

Levy, D. (2005). *Robots Unlimited: Life in a Virtual Age.* New York: A K Peters/CRC Press. https://doi.org/10.1201/b10697

Levy, D. (2009). *Love and Sex with Robots: The Evolution of Human-Robot Relationships.* Harper Collins.

Lugano, G., Hudák, M., Ivančo, M., & Loveček, T. (2019). From the Mind to the Cloud: Personal Data in the Age of the Internet of Things. In Y. Zhou & M. H. Fischer (Eds.), *AI Love You.* Springer Nature Switzerland.

McClelland, T. R. (2017, Summer and Autumn). Confronting Emerging New Technology: The Case of the Sexbots. *The Journal of Mind and Behavior, 38*(3), 247–270.

Nye, D. E. (2006). *Technology Matters: Questions to Live With.* Cambridge: MIT Press

Nyholm, S. (2020). *Robots: ethics, Agency, and Anthropomorphism.* Rowman & Littlefield.

Scheutz, M. (2011). The Inherent Dangers of Unidirectional Emotional Bonds Between Humans and Social Robots. In P. Lin, K. Abney, & G. A. Bekey (Eds.), *Robot Ethics: The Ethical and Social Implications of Robotics* (pp. 187–202). MIT Press.

Sharkey, N., A. Van Wynsberghe, S. Robbins, & E. Hancock. (2017). Our Sexual Future with Robots: a Foundation for Responsible Robotics Consultation Report. Retrieved March 13, 2021, from https://responsible-robotics-myxf6pn3xr.netdna-ssl.com/wp-content/uploads/2017/11/FRR-Consultation-Report-Our-Sexual-Future-with-robots-1-1.pdf

Sullins, J. P. (2012, January). Robots, Love, and Sex: The Ethics of Building a Love Machine. *IEEE Transactions on Affective Computing, 3*(4), 398–409. https://doi.org/10.1109/T-AFFC.2012.31.S2CID253828

Tibbals, C. (2016). Sex Robots Misquoting & Reason #74,193 I Only Do Written Interviews | Dr. Chauntelle Tibbals. Retrieved September 15, 2020.

Turkle, S. (2004). Whither Psychoanalysis in the Computer Culture? *Psychoanalytic Psychology, 21*(1), 16–30.

Van de Poel, I. (2016). An Ethical Framework for Evaluating Experimental Technology. *Science and Engineering Ethics, 22*(3), 667–686.

Van de Poel, I. (2020). Three Philosophical Perspectives on the Relation Between Technology and Society, and How They Affect the Current Debate About Artificial Intelligence. *Human Affairs, 30*, 499–511.

Van de Poel, I., & Royakkers, L. (2011). *Ethics, Technology, and Engineering : An Introduction*. Wiley-Blackwell. http://search.ebscohost.com/login.aspx?direct=true&scope=site&db=nlebk&db=nlabk&AN=510116

Van Grunsven, J. (2020). Perceiving 'Other' Minds: Autism, 4E Cognition, and the Idea of Neurodiversity. *The Journal of Consciousness Studies, 27*(7–8), 115–143.

Van Grunsven, J., & van Wynsberghe, A. (2019). A Semblance of Aliveness: How the Peculiar Embodiment of Sex Robots Will Matter. *Techne: Research in Philosophy and Technology, 23*(3), 290–317. https://doi.org/10.5840/techne20191125105

Wachter, S. (2018). Normative Challenges of Identification in the Internet of Things: Privacy, Profiling, Discrimination, and the GDPR. *Computer Law & Security Review, 34*, 436–449.

Weinberg, A. M. (1991). Can Technology Replace Social Engineering? In W. B. Thompson (Ed.), *Controlling Technology: Contemporary Issues* (pp. 41–48). Prometheus Books.

Whitby, B. (2011). Do You Want a Robot Lover? In P. Lin, K. Abney, & G. A. Bekey (Eds.), *Robot Ethics: The Ethical and Social Implications of Robotics. Intelligent Robotics and Autonomous Agents* (pp. 233–249). MIT Press. ISBN 9780262016667.

The Right and Unfair Aspects of Artificial Womb Technology

Vera Tripodi

1 Prologue

Ectogenesis (the gestation of a foetus outside the womb) is an important gender issue that is also relevant to feminism. Its importance is obvious for at least two reasons. First, the artificial womb is a technology that might make gender characteristics less relevant to reproduction or parenting and replace the biological practice of gestation, a capacity that only women (although not all) have. It would thus remove the exclusive biological connection between female sex and pregnancy by disconnecting motherhood from female biological physicality. Second, artificial womb might allow men to gestate and become fathers without the help of a woman surrogate. Thus, women's pregnancy would no longer be necessary (Murphy, 1989; Pence, 2006). In short, it is a technology that could increase or decrease—depending on the point of view—reproductive freedom and autonomy for women (Johnstone, 2010). But, what are

V. Tripodi (✉)
University of Milan, Milan, Italy
e-mail: vera.tripodi@unimi.it

E. Terrone, V. Tripodi (eds.), *Being and Value in Technology*,
https://doi.org/10.1007/978-3-030-88793-3_5

the conceptual and ethical implications of artificial-womb technology (AWT) on humans and in particular on women?

In *The Second Sex*, Simone de Beauvoir writes:

> Artificial insemination completes the evolutionary advance that will enable humanity to master the reproductive function. These changes are of tremendous importance for woman in particular; she can reduce the number of her pregnancies and make them a rationally integral part of her life, instead of being their slave. During the nineteenth century woman in her turn emancipated herself from nature; she gained mastery of her own body. Now protected in large part from the slavery of reproduction, she is in a position to assume the economic role that is offered her and will assure her of complete independence. (De Beauvoir, 2010, p. 143)

Much has changed since de Beauvoir wrote these lines. Our reproductive methods have developed, and science has questioned the limits of human reproduction, limits that at the time de Beauvoir wrote, seemed natural and obvious (Arditti et al., 1984; Grossman, 1971). Medical technology has made great progresses in assisted fertilization over the past forty years, making it possible for children who otherwise, with natural methods, would never have been able to come into the world to be born. Some scientists believe that the artificial womb will be the new frontier of human reproduction—a sort of very sophisticated incubator in which embryos can develop without need for the woman's body (Bulletti et al., 1986, 2011; Romanis, 2018). It is thus a technology that could help those suffering from recurrent miscarriages to have offspring, save premature foetuses (Moore, 2017), and replace uterus transplants. But according to some authors, the development of AWT might drastically change the naturalness of pregnancy (Burley, 1998; Murphy, 2012).

In 2017 and 2019, in the USA and Australia respectively, researchers successfully developed and tested animal artificial womb devices, called the "Biobag" (Mullin, 2017; Partridge et al., 2017; Usuda et al., 2019). Some scientists (Bulletti et al., 2011) are keen to argue that AWT might be an innovative, beneficial treatment for humans. This technology, some experts say (Partridge et al., 2017; Usuda et al., 2019), might present important therapeutic aids and benefit a great number of human beings, particularly babies and women. First, it could significantly reduce morbidity rates among preterm babies. Thus, it might be an alternative to neonatal intensive care and reduce deaths due to prematurity. Second, artificial wombs could mitigate the physical constrictions currently faced by some women. Women may use an artificial womb if becoming pregnant is deemed too risky or they may transfer their foetus to an artificial womb if risk increases as pregnancy progresses. Aside from uterine transplants and surrogacy, AWT might even be an alternative fertility treatment for women unable to carry a full pregnancy or for men who could then have babies without the help or assistance of women for a long nine months. In addition, for a single man or a couple of two men, ectogenesis might enable the possibility of becoming a parent of one's own child.

One of the central issues in feminist ethics and bioethics regarding artificial wombs, on which this contribution will be focusing, is whether AWT would be a tool for women's liberation and emancipation. There is a genuine philosophical disagreement among feminists about whether AWT could have also important social benefits for women (Kendal, 2015; Smajdor, 2007, 2012, 2016). According to some feminists, artificial wombs liberate women from patriarchal models and the heavily gendered process of reproduction (Takala, 2009). Free from their reproductive duties, women might better pursue their interests and desires. By contrast, some feminists hold that the diminishing of the maternal-foetal relationship is unethical and doubt that ectogenesis would resolve the socioeconomic inequalities between sexes and genders.

In this contribution, I defend the view of AWT as free reproductive choice and argue that ectogenesis technology should become a morally acceptable option without making natural pregnancy obsolete. The chapter is divided into two parts. In the first part, I point out arguments in favour and against ectogenesis. In the second part, I show how the artificial womb can be seen as providing a free choice for women (a technology that might be used also by women who are not infertile and for whom pregnancy is not a risk) and as a tool for partially ending the unequal division of reproductive labour.

2 Feminist Arguments in Favour of Ectogenesis and Artificial Wombs

Preserving generation and reproduction is necessary for the perpetuation of all species (for both male and female members of each species). Nevertheless, when it comes to human beings, the greatest responsibility for the perpetuation of our species is on women's shoulders. With respect to reproductive function, female individuality is sacrificed in favour of the species. For humans, gestation is the period in which the embryo develops in a woman's body (or in a biologically female subject). So, gestation and pregnancy are physical phenomena with significant mental impacts that involve only women or female human beings, both in heterosexual and lesbian relationships (in the latter case, thanks to sperm donation and assisted fertilization). According to Shulamith Firestone, the natural method of continuing our species is the basic cause of women's inequality. More precisely, in *The Dialectic of Sex* (1970), Firestone maintains that pregnancy is barbaric and harmful to women. This seems to be in keeping with what de Beauvoir thinks about gestation and pregnancy, namely that it is "a fatiguing task of no individual benefit to the woman but on the contrary demanding heavy sacrifices" (Firestone, 1970, p. 57). What characterizes human female gestation is a long and often difficult pregnancy, painful and sometimes dangerous childbirth, illnesses, and unexpected symptoms and complications. During pregnancy, women may in fact experience serious unexpected adverse medical events or dangerous disorders, metabolic over-activity, or fatigue and anxiety. Many may also suffer from loss of appetite and vomiting or lack of phosphorus, calcium, and iron, and recouping these losses after childbirth can be very difficult. Moreover, many problems or complications can develop during labour and delivery: childbirth itself is painful and dangerous for women; during delivery, women may develop a chronic illness or die. Breastfeeding and taking care of children is also a tiring service that might be crushing if the woman is compelled to nurse and left without assistance. Furthermore, pregnancy and childbirth decrease a woman's capacity for employment and may at times make her wholly dependant upon others, such as a partner, family members, or friends.

One might object that women experience pregnancy and motherhood very differently: some women experience both with resignation, others with satisfaction or enthusiasm. There are also women who enjoy the pleasure of breastfeeding. Nonetheless, according to Firestone and de Beauvoir, pregnancy is above all a drama for a woman. A pregnant woman, de Beauvoir maintains (2010), is tossed and driven, she feels the foetus as "a part of her body, and it is a parasite that feeds on it" (p. 476), as such, she possesses it and she is possessed by it. De Beauvoir believes that pregnancy is not the essence of being female and that it is never an enrichment for a woman, but rather, that it decreases her ego. The woman lives through pregnancy with an attitude of ambivalence because she feels it as both an enrichment and a mutilation. Ensnared by nature, de Beauvoir writes, the pregnant woman "is plant and animal, a storehouse of colloids, an incubator, an egg" (De Beauvoir, 2010, p. 478). However, de Beauvoir highlights that the social context ensures that "the biological condition of woman does constitute a handicap" (De Beauvoir, 2010, p. 331). As matter of fact, the passivity conventionally attributed to a woman and her socially subordinate condition have nothing to do with the anatomy of her sexual organs or her biological traits. Rather, it is "a destiny imposed upon her [...] by society" (De Beauvoir, 2010, p. 285). Therefore, the biological data are not to be taken as a "rigid destiny" and the biological facts "do not condemn her to remain in this subordinate role for ever" (De Beauvoir, 2010, p. 61).

In line with Firestone and de Beauvoir, in more recent times, some feminists (e.g., Kendal, 2015) have argued that ectogenesis is morally appreciable and desirable for women for at least three reasons. First, ectogenesis might free women from the unjust and painful burden of pregnancy, the tyranny of gestation, or unwanted aspects of pregnancy. Ectogenesis makes it possible for women to have children without becoming pregnant and could be used by women who are at high risk for complications during pregnancy without turning to a surrogacy. For those who need abortion when their life is in danger during pregnancy, an artificial womb would offer a way to avoid having to terminate foetal development.

Second, ectogenesis is a technology that might expand women's reproductive capabilities. As noted above, nature causes women's (but not men's) reproductive capacity to decrease and eventually cease with age. But this fact need not be accepted as an inexorable destiny. Hence, AWT might remove age limits to gestation: women might postpone motherhood (freezing their eggs) without panicking about their "biological clock", plan to become mothers later, or focus—without renouncing having children—on other aspects of their work and social life. In patriarchal societies, femininity is reduced to motherhood, and woman to the condition of mother (Wishart, 1982). Patriarchy has used (natural) motherhood as an instrument for the subjection and control of women by reducing and forcing them towards the domestic, private, and invisible sphere. As an obligation, motherhood was (and in many social contexts still is) an obstacle to women's aspirations and autonomy (Badinter, 1981). In this picture, the liberation and emancipation of women therefore means leaving home, achieving economic independence, and putting childcaring aside. Thus, ectogenesis can be an important technological tool for creating social equality in becoming parents, which nature seems to deny, between men and women.

Third, AWT it is a technology that might promote gender equality in reproduction. Thanks to ectogenesis, the woman's contribution to reproduction will be similar to the man's. To birth their babies, in the future women and men will need only to provide or donate their gametes (eggs or sperm). As a consequence, like men, women could maintain their normal social and working life for the nine months necessary for a foetus to develop and grow.

Moreover, for those who find gestational surrogacy and the supposed exploitation of women's bodies ethically problematic, ectogenesis might allow men the possibility of becoming parents without commodifying the female body. The artificial uterus would undercut and override many of the objections feminists have against surrogacy, and many gay couples or single intended fathers might consider finding and select an egg donor and then choosing ectogenesis as a way to start or add to their family.

In short, some feminists claim, ectogenesis might enhance women's reproductive rights, contribute to gender equality, and reduce the huge amount of reproductive work usually carried out by women.

3 Feminist Arguments Against Ectogenesis and Artificial Wombs

Many of the feminist arguments against ectogenesis are based on the following three beliefs: i) medical technologies are often instruments from the male domain that expose a cause-and-effect relationship between biotechnologies and the commercialization of biological reproduction; ii) medical technologies are often used as a tool for increasing the subordination of women; iii) the use of biotechnology like AWT may not be in the best interests of the future child.

In what follows, I will not discuss the last point, namely, issues regarding the rights of future children and their supposed right to be birthed by a human mother. I think this is a significant ethical issue, although "the best interests for the chid" can be an ambiguous phrase and there are cases in which the biological pregnancy is not in the child's interests. Here, I would simply say that the worry about these issues has to do with the fact that, at the time, we cannot guarantee that the future child born through ectogenesis would develop normally and not be physically or mentally disadvantaged. But, I would underline, the goal of ectogenesis is to produce an infant indistinguishable in health from an infant born from a woman, and in this chapter I am considering a future scenario in which ectogenesis is established as a safe practice. Thus, in what follows, I will limit my focus to (i) and (ii) above. Let us start by discussing point (i).

According to some feminists (Oakley, 1984), as has happened with other emerging technologies, ectogenesis will be a new tool of male domination; it will be used to benefit men and devalue women. More precisely, ectogenesis raises the problem of medical control over women (their bodies, their maternal practices, their lives) and the male control of reproductive technologies. As long as biological technologies are in the hands of men, women certainly won't—this is the conviction—benefit from them. This is in part because these technologies are connected to mechanisms that involve the surveillance of motherhood and thus implicitly the social control of women.

A further worry is that through ectogenesis, men aim to divorce women from the experience of gestation and pregnancy because of their fear of the procreative power women hold (Rich, 1977). The birth experience is, as Rich stresses, a ritual that women have historically shared. This traditional ritual has been broken by the intervention of medical technologies. The experience of pregnancy (and menstruation and menopause) is the expression of the universal relationship of women to a new life. This experience gives women a deep bond with other mothers and a sense of continuity, forming a continuous line of women from antiquity to present day. As some feminist theorists have stressed (see Rich, 1977), the experience of pregnancy is not an impediment to gender equality and it is itself something positive, worthwhile, wonderful, and powerful. Moreover, many women want to become a mother, have the experience of pregnancy, and find mothering gratifying. But with uptake of new technologies, men become more involved in birthing, which can be problematic. In fact, the history of medicine provides many examples of reproductive technology being used for controlling (and not liberating) women, as well as the usurpation of the birth process by the male-domain (Amoretti & Vassallo, 2017; Kukla, 2005). The term "male control" here does not mean control by an individual man. Rather, it refers to the control that benefits—most of the time—more men than women (Albury, 1984). We are in a society in which there is a network of power relations that ensures the success of male individuals by reinforcing a certain ideal of masculinity and patriarchy. The ethic of control in medicine has encouraged research on reproductive technologies. But many (male) scientists are reluctant to consider the social implications of their work, address the needs of women, and expand research in reproductive technologies while engaging community debate. Moreover, as some feminists have highlighted, many women will not have access to these technologies, perhaps in part because women can be also differentially disadvantaged because of their race and class: women are often divided into 'worthy' and 'unworthy' mothers on the basis of their race and class membership. So, new reproductive technologies do not necessarily give all women greater control over their lives. As some feminists and activists (including Andrea Dworkin and Janice Raymond) point out, the reproductive technologies currently available are not moving women towards emancipation and are part of the male-dominated project to deprive women of their reproductive powers.

In addition, on this perspective, ectogenesis is unnatural, that is, it is a technology the job of which is to alter our basic biological condition and violate natural laws. Alongside other emerging reproductive technologies, some feminists argue, ectogenesis is problematic insofar as it threatens to disrupt the natural link between reproduction and mothering (Corea, 1985a, 1985b; Corea et al., 1985; Raymond, 1987, Rowland, 1987). The power of this technology is seen as a "threat" to women and, as a consequence, it raises the issue of the "limit" that should be placed on technological development and innovation: technology allows for the manipulation of life and human life in particular. Here the worry is that not being moved by pure intellectual interests, the "techno-science" would end up being subjected to the technological imperative according to which "everything that is physically and technically possible to do" can be done. But some authors (Jonas, 1984) emphasize that not everything that is technically possible or feasible is ethically acceptable. They argue that the natural limit, inscribed in our body, should not be violated. This argument seems to be linked to the moral paradigm of "the Sanctity of Life" (Kuhse, 1987; Singer, 1994; Singer & Wells, 1984; Singer & Wells, 2006; Tonti-Filippini, 2003; Warren, 1986) and to a particular philosophical anthropology, that is, a conservative conception of the human being and the world according to which the natural limit must be absolutely respected because in that limit lies the supposed wisdom of nature, a wisdom that may not be immediately grasped by our (limited) individual reason, but that exists and reveals itself over time.

Let us now examine point (ii). Feminist theorists who think that AWT is a technology that downgrades the female biological capacity see, as mentioned above, motherhood in a positive light. In *Of Woman Born* (1977), Adrienne Rich distinguishes the institution of forced motherhood, as imposed by the patriarchy, and women's potential relationship with the experience of motherhood (as opposed to forced motherhood). According to Rich, we should eliminate the "institution of motherhood" and place motherhood outside the sphere of the patriarchy. This does not mean abolishing motherhood. On this view, the reproductive capacity of the female body is not a barrier to emancipation. Rather, pregnancy is understood as a resource and not a destiny.

Consequently, women's liberation goes through the defence of female sexual, reproductive, and maternal potential. In contrast to de Beauvoir's view regarding the liberation of women from reproductive tasks, radical feminism argues that attention to female sexuality (in particular to woman's ability to give birth to a new life) is instead the starting point for the affirmation of the power and value of the female body (O'Brien, 1981). Radical feminism thus celebrates the power of female sexuality as what allows one to escape male domination and submission. It emphasizes the maternal body as the source of positive values against patriarchal norms (Lorde, 1984; Rich, 1977) and praises the typically feminine qualities, particularly care and intersubjectivity (as opposed to the masculine-related qualities of autonomy and duty).

According to this perspective, it is her reproductive capacity that also makes a woman more inclined than a man to maintain a pacifist disposition or an anti-militarist vision, and to be more sensitive to environmental and ecological issues (Griffin, 1978). As Adriana Cavarero (1995, 2000) points out, the body is the basis of the first and most fundamental relationship, indispensable for the creation of a new life, that of the child who is born from the mother's body, or rather from a female sexed body. This reflection has allowed feminists to rethink and reconstruct the stereotype of the woman mother, purifying it of the servile aspect inherited from patriarchal thought.

Furthermore, to remind us that we are also bodies and not only minds, the idea (and theories) of sexual difference is based on an ethics of exposure, of need, of inclination towards the other, and on a subject open to injury because the subject, as embodied in a body, is always vulnerable. The self is necessarily inclined towards others and is co-constructed with others (Cavarero, 2000). On Cavarero's view (Cavarero, 1995), in Western metaphysics, there is a lack of attention to the fact that we are all born from a woman. In this philosophical tradition, the concern is on death rather than birth.

Alison Stone (2011) has explored the maternal body and proposes models of subjectivity immersed in relations of intimacy and dependence. This form of subjectivity can be explained in terms of how the mother typically reproduces her history of bodily relations with her own mother with her child. This leads to a maternal and cyclical form of lived time (Stone, 2011). And, following Julia Kristeva's theory of the semiotic and the symbolic (1980), the dimension of the semiotic finds its origin in the bond with the maternal body and in the mother's relationship with the child. The semiotic order consists of signs and images and corresponds to a pre-discursive phase. It therefore precedes the separation from the mother's body and the articulation of a form of linguistic-symbolic communication. Because of the values it expresses, therefore, the female body should be a source of pride and not, as de Beauvoir sometimes seems to argue, of shame. Motherhood and pregnancy are hence the range of potential empowerment.

There are two other worries about AWT for feminist theorists arguing against ectogenesis. The first is linked to the issue of abortion. In the future, when ectogenesis will be a safe and widely available practice, women might be forced to reproduce via artificial womb, or it could be a moral obligation for women not to have a child in a natural way if it is considered riskier to do so. Thus, some authors think (Langford, 2008), ectogenesis might limit women's reproductive liberties and women who want to have a natural pregnancy would be considered irresponsible. Women who have, for health reasons, a high risk of having an abortion or complications in pregnancy would be discouraged from pursuing natural gestation. Thus, ectogenesis could also be used as a means of resolving the abortion debate (Brassington, 2009; Cannold, 1995; Steiger, 2010). According to the feminist argument for the freedom to choose abortion, women have the right to terminate a pregnancy and should not be pregnant against their will. But is this the case of AWT (in which women's body is not involved)? Would a woman still have a right to abort in the case of an unwanted ectogenic pregnancy? In the case of AWT, does the foetus belong in women's body? Individual freedom implies having a right over one's own body and a pregnant woman has the right to control her body, to choose to not be pregnant. But AWT eliminates the need for women to bear children. With ectogenesis, a woman could in theory abort the foetus without killing it; she could terminate her pregnancy without terminating the foetus's life. Some feminists thus highlight that AWT might endanger and challenge women's abortion rights

(Overall, 2015; Räsänen, 2017). However, some theorists have replied by asking how, if the bond with the mother is so intimate and profound, abortion can be justified from a feminist perspective at all (Satz, 1992). Other authors object that AWT is open to criticism without calling into question the intimacy of the relationship with the foetus.

The second worry is that we do not have any guarantee that women, free from the unjust burden of pregnancy, will equally participate in the social and political sphere because women's subordination is a more complex phenomenon rooted in customs, culture, and legal restraints. AWT might not be sufficient to end women's social and political oppression.

In sum, according to some feminists, ectogenesis might be a new form of pressure, oppression, and devaluation of women; the artificial womb might be a harm and danger for women. It is a technology that might limit or decrease women's reproductive freedom and autonomy, and their control to exercise reproductive rights, especially in patriarchal societies.

4 Ectogenesis as Free Reproductive Choice

The progress achieved in the biomedical and healthcare fields has called into question some biological limits that until a few years ago seemed natural, obvious, and solid. Assistive reproductive technologies today help people who want to be parents with conceiving children by intervening in the vital processes and questioning the millenary sacral dimension of life (Buuck, 1977; Chavatte-Palmer et al., 2012). Assisted fertilization involves a profound change (Donchin, 1986), not only in reproductive methods, but also in the social structure of the traditional family. For example, it opens the possibility for same-sex couples to reproduce and have children. To assess whether the new reproductive technologies (like AWT) are ethically acceptable, one might say that we should answer two different questions with the knowledge that our answers to those questions might conflict each other: (1) Why should technical interventions that increase life opportunities and fulfil a good purpose (i.e., allowing people to have children) be considered immoral? (2) Why shouldn't we go beyond the pillars of Hercules of natural reproduction in a woman's body? Why not force nature a little where this forcing promises good consequences?

Here, however, a different order of problems arises. The first regards the interests and rights of individuals or couples who want to reproduce. Who, for example, is entitled to these reproductive technologies, and under which conditions? Second, questions arise regarding the interests and rights of future generations. For example, how might being born through laboratory methods and without sexual intercourse affect the physical and mental development of these individuals? What are the possible harms for the future child and how can these balance with the interests of the parents who choose to reproduce in an unconventional way? Third, we should consider the moral status of the entity on which we intervene with these technologies. What is the moral status of the individual grown in artificial womb? What does an artificial womb mean for the freedom to abort or end that individual's life? The implications of the new forms of reproduction are vast and not yet fully explored.

Now, as we saw, new reproductive technologies are principally designed to treat infertility or induce pregnancy, and ectogenesis is currently a technology being studied to improve neonatal care and assist infertility research (Gelfand, 2006, Pfeffer & Woollett, 1983). In what follows, I do not focus on these aspects of ectogenesis. Rather, the issue I am interested in is whether there are good reasons for considering ectogenesis as a woman's choice even in the absence of health problems; whether it could be a morally acceptable choice for a woman to prefer ectogenesis even when she can reproduce (without risks) naturally and without infertility problems; whether ectogenesis makes pregnancy obsolete and an inferior act, and women should be free from pregnancy. My view is that the availability of ectogenesis does not imply that women ought to refrain from pregnancy if they want a baby and to become a mother. What I am going to argue in this last part of the chapter is that women should be free to choose ectogenesis (if they want to have a baby but not a pregnancy) and that this affirms women's reproductive rights.

Ectogenesis also offers an alternative for women who are capable of pregnancy or for whom pregnancy is not risky. First, this possibility would not make motherhood obsolete. Wishing for a baby, or to be a mother, is different from wishing to be pregnant. Properly speaking, ectogenesis would not in fact be a form of surrogacy of motherhood. Rather, it would be a surrogacy of pregnancy. In fact, ectogenesis might not undermine the roles that belong to the woman who intends to play the social role of mother, namely the specific social and emotional responsibilities towards a son or daughter, such as the obligation to economically support, instruct, and educate him or her. In defence of the free choice to use AWT, my starting point is, as previously said, that the goal of ectogenesis is to give birth to a healthy infant, and its desirability is predicated on the assumption that this technology would not produce foetal harm.

With this premise in mind, I suggest that ectogenesis:

(i) should be considered a morally permissible medical practice;
(ii) should be allowed both to couples (heterosexual or of the same sex) and to singles (regardless of gender or sex) (Kimberly et al., 2020);
(iii) should be considered a legitimate path to "found a family" and "have children".
(iv) should be allowed for both women who have health problems and those who don't.

A woman's choice for ectogenesis is, I argue, within her ability to self-determine. At the time such as decision is made, we cannot have any guarantee that the future child will develop normally or that the child will not be permanently (physically or mentally) disadvantaged. As previously stated, there is currently no scientific evidence that ectogenesis involves damage for others or for the unborn child. But, for the sake of the argument, let us suppose there are no studies that offer good reasons to argue that those born through ectogenesis risk developing physical anomalies and pathologies. And let us suppose that there is no evidence in the scientific literature that leads to the belief that those born through this practice are destined for a worse or unworthy life than those born through more common and conventional reproductive practices. If this is the case, we might say that—in the absence of proven evidence that an action or medical practice involves the violation of the rights or freedoms of others, that is, in the absence of evidence of damage to third parties—practice can be considered morally permissible.

The reasons why a woman might choose ectogenesis for having a baby can be very different. As we saw, there may be clinical reasons, which depend for example on a woman's inability to start or carry on a pregnancy. For others, gestation may be the only way to have a child. This may be the case, for example, for women affected by the Mayer-Rokitansky-Küster-Hauser (MRKH) syndrome (also known as Mayer-Rokitansky syndrome or Rokitansky-Küster-Hauser syndrome) and who are therefore affected by a congenital absence of the uterus, for women who have undergone a hysterectomy or ovarian removal, for those who have already undergone unsuccessful artificial insemination cycles, or for those whose health status is considered incompatible with pregnancy.

Alongside health, there may also be personal reasons for a woman to prefer ectogenesis (Simonstein & Mashiach-Eizenberg, 2009). This could be due to her personal, work, or economic circumstances. For example, ectogenesis might avoid interruptions related to career (say, for an actress, dancer, or a professional athlete). Moreover, a woman might have professional or care commitments which she cannot avoid (a position of responsibility or a disabled family member) and which she cannot delegate. Additionally, a woman might have deep fear related to previous pregnancies, or difficulties managing the emotional and physical stress of pregnancy. The personal reasons that can push a woman who wants a baby to resort to an artificial uterus are—in my view—no less relevant than the clinical ones.

Ectogenesis can be inscribed, I think, in the recognition of motherhood as one of the deepest human desires and one of the most intimate aspirations of human beings. I agree with those who argue that it does not matter how you become a mother to be a good mother, that is to say, to be caring and responsible towards your children or to take the utmost account of your moral duty to ensure the best possible interests for their growth and realization. Furthermore, if there is a moral duty to make the world a better place and make people happy, then allowing those who cannot have children of their own means or who find pregnancy a difficult or unpleasant experience to resort to AWT means benefiting many people and contributing to their happiness. Thus, it might be argued that it is even morally permissible for a woman who wishes to have a genetic lineage but at the same time avoid (and not desire) a pregnancy, to use this technology. Being a good parent (in this case a mother) does not depend so much on participating physically in the biological event of conceiving or on gestation, but on assuming the responsibility of giving birth to an individual without his or her consent.

Moreover, as I mentioned above, the change in the case of ectogenesis is mostly read as a danger for children. One might say that a child born through this medical practice cannot be protected by her or his mother who was so selfish that she was willing to do anything to satisfy the desire to have a child of her own. One might think that those women who resort to ectogenesis are moved by a strong selfish desire, that of having a child of their own without going through "the joys and sorrows" of pregnancy. Is the ego, in this case, really the driving force behind satisfying the desire for motherhood? What role does selfishness play in desiring a mothering experience that passes through AWT? To analyse this point, let us consider a general reflection on the relationship between innovation and desire.

Technologies (especially biomedical) usually broaden our horizons and offer new possibilities (Finn & Isaac, 2021; Kingma & Finn, 2020). Accepting these possibilities involves, mostly, a revision of our cultural schemes or paradigms with which we look at and classify the world, as well as a modification of some social norms. The same goes for the new technologies of assisted reproduction. These technologies allow us to control human reproduction and offer new possibilities and ways to realize the desire to have a child of our own. This possibility has produced a radical change in our culture. Therefore, many questions concerning possible ways in which to reproduce can no longer be analysed or evaluated, even from a moral point of view, using traditional schemes and paradigms. With the biomedical revolution—which made it possible to manipulate the organic human life—a new chapter in history has opened for humanity.

To understand the importance of this change, we should go further than the traditional paradigm of the "Sanctity of Life". This paradigm is opposed to that of the "Quality of Life". According to the first paradigm, human life is intangible and cannot ever be violated or manipulated (on this perspective, abortion and assisted human reproduction are both unethical and should be prohibited). In the "Quality of Life" paradigm, the central value is quality of life understood as well-being and respect for people's autonomy. Ectogenesis, together with the other ways that technology allows one to become a mother, allows for the expression of new desires, for example, having a child without a pregnancy where pregnancy is understood as an unpleasant experience, which once did not exist because science did not offer these possibilities and did not open to these scenarios. It is therefore a duty, I suggest, not to hinder innovations that can offer conditions of happiness to people.

Reproductive technologies raise many ethical questions (Warren, 1989; Wolfe, 2008), but in responding to them, we must not be afraid of the new and the radical changes that technological innovation (in the field of medicine and human reproduction in particular) could bring about. The desire to have a child of one's own, that is expressed through recourse to AWT, is therefore no more selfish than it is to have a child of one's own through more conventional methods.

It should be emphasized that the desire for parenthood—even that which is achieved through AWT—is included among human needs (regardless of one's sexual orientation). It is therefore not a question of a frivolous or capricious desire. Rather, the deep desire to have a child of one's own should not be underestimated or trivialized. Certainly, the satisfaction of this desire cannot override everything. There are limits to the realization of this desire that must be respected, such as the protection of the child who we intentionally decide to bring into the world. However, it should be stressed that conscious parenthood is an important life project, even when achieved with the help of technology.

In sum, ectogenesis could be an ethically unproblematic alternative not only for women who are incapable of pregnancy or for whom pregnancy is not recommended on medical grounds, but also for those who want children without becoming pregnant. It does not follow from this, however, that there should be a moral obligation to refrain from having a child naturally. Rather, ectogenesis should be a free choice for women and non-clinical reasons should also be considered good reasons for choosing this practice.

References

Albury, R. (1984, Spring). Reproductive Technology and Feminism. *Australian Left Review, 89*, 46–55.

Amoretti, C., & Vassallo, N. (2017). Philosophy of Sex and Gender in Gender Medicine. *Special issue of Topoi, 36*(3), 473–559.

Arditti, R., Duelli, K. R., & Minden, S. (Eds.). (1984). *Test-Tube Women: What Future for Motherhood?* Pandora Press.

Badinter, E. (1981). *The Myth of Motherhood: An Historical View of the Maternal Instinct*. Souvenir Press.

Brassington, I. (2009). The Glass Womb. In F. Simonstein (Ed.), *Reprogen-Ethics and the Future of Gender*. Springer.

Bulletti, C., Jasonni, V. M., Lubicz, S., et al. (1986). Extracorporeal Perfusion of the Human Uterus. *American Journal of Obstetrics and Gynecology, 154*, 683–688.

Bulletti, C., et al. (2011). The Artificial Womb. *Annals of the New York Academy of Sciences, 1221*, 127.

Burley, J. (1998). The Price of Eggs: Who Should Bear the Cost of Fertility Treatments? In J. Harris & S. Holm (Eds.), *The Future of Human Reproduction: Ethics, Choice, and Regulation* (pp. 127–149). Clarendon Press.

Buuck, J. (1977). Ethics of Reproductive Engineering. *Perspectives, 3*(9), 545–547.

Cannold, L. (1995). Women, Ectogenesis and Ethical Theory. *Journal of Applied Philosophy, 12*(1), 55–64.

Cavarero, A. (1995). *In Spite of Plato*. Polity Press.

Cavarero, A. (2000). *Relating Narratives: Storytelling and Selfhood*. Routledge.

Chavatte-Palmer, P., Levy, R., & Boileau, P. (2012). Reproduction without a uterus: State of the art of ectogenesis. *Gynecologie Obstetrique & Fertility, 40*, 695–697.

Corea, G. (1985a). How the New Reproductive Technologies Could Be Used to Apply the Brothel Model of Social Control Over Women. *Women's Studies International Forum, 8*(4), 299–305.

Corea, G. (1985b). *The Mother Machine: From Artificial Insemination to Artificial Wombs*. Harper & Row.

Corea, G., et al. (Eds.). (1985). *Man-made Women: How the New Reproductive Technologies Affect Wome*. Hutchinson.

De Beauvoir, S. (2010). *The Second Sex*. (C. Borde & S. Malovany-Chevallier, Trans.) Alfred A. Knopf.

Donchin, A. (1986). The Future of Mothering: Reproductive Technology and Feminist Theory. *Hypatia, 1*(2), 121–138.

Finn, S., & Isaac, S. (2021). Evaluating Ectogenesis Via the Metaphysics of Pregnancy. In R. Davis-Floyd (Ed.), *Birthing Techno-Sapiens: Human-Technology, Co-Evolution, and the Future of Reproduction*. Routledge. chapter 8.

Firestone, S. (1970). *The Dialectic of Sex: The Case for Feminist Revolution*. William Morrow and Company.

Gelfand, S. (2006). *Ectogenesis and the Ethics of Care*. In S. Gelfand & R. Shook (Eds.), *Ectogenesis: Artificial Womb Technology and the Future of Human Reproduction*. Rodopi.

Griffin, S. (1978). *Woman and Nature: The Roaring Inside Her*. Harper and Row.

Grossman, E. (1971). The Obsolescent Mother: A Scenario. *Atlantic, 227*, 39–50.

Johnstone, M. (2010). Ethics and Ectogenesis. *Australian Nursing Journal, 33*(11), 33.

Jonas, H. (1984). *The Imperative of Responsibility: In Search of Ethics for the Technological Age*. The University of Chicago Press.

Kendal, E. (2015). *Equal Opportunity and the Case for State Sponsored Ectogenesis*. Palgrave Macmillan.

Kimberly, L. L., Sutter, M. E., & Quinn, G. P. (2020). Equitable Access to Ectogenesis for Sexual and Gender Minorities. *Bioethics*. https://doi.org/10.1111/bioe.12723

Kingma, E., & Finn, S. (2020). Neonatal Incubator or Artificial Womb? Distinguishing Ectogestation and Ectogenesis Using the Metaphysics of Pregnancy. *Bioethics*. https://doi.org/10.1111/bioe.12733

Kristeva, J. (1980). *Desire in Language. A Semantic Approach to Literature and Art*. Columbia University Press.

Kuhse, H. (1987). *The Sanctity-of-Life Doctrine in Medicine. A Critique*. Clarendon.

Kukla, R. (2005). *Mass Hysteria: Medicine, Culture, and Mothers' Bodies*. Rowman & Littlefield Publishers.

Langford, S. (2008). An End to Abortion? A Feminist Critique of the "Ectogenetic Solution" to Abortion. *Women's Studies International Forum, 31*(4), 263–269.

Lorde, A. (1984). The Master's Tools Will Never Dismantle the Master's House. In Id. (Ed.), *Sister Outsider* (pp. 110–114). The Crossing Press, Freedom.

Moore, A. (2017, April 25). A Unique Womb-Like Device Could Reduce Mortality and Disability for Extremely Premature Babies. *CHOP News*.

https://www.chop.edu/news/unique-womb-device-could-reduce-mortality-and-disability-extremely-premature-babies

Mullin, E. (2017, April 25). Animals Set Survival Record Inside Artificial Womb. *MIT Technology Review*. https://www.technologyreview.com/s/604261/animals-set-survival-record-inside-artificial-womb/

Murphy, J. (1989). Is Pregnancy Necessary? Feminist Concerns about Ectogenesis. *Hypatia, 4*(3), 66–84.

Murphy, T. F. (2012). Research Priorities and the Future of Pregnancy. *Cambridge Quarterly of Healthcare Ethics, 21*(1), 78–89.

O'Brien, M. (1981). *The Politics of Reproduction*. Routledge & Kegan.

Oakley, A. (1984). *The Captured Womb: A History of the Medical Care of Pregnant Women*. Basil Blackwell.

Overall, C. (2015). Rethinking Abortion, Ectogenesis, and Fetal Death. *Journal of Social Philosophy, 46*, 126–140.

Partridge, E., Davey, M., Hornick, M., et al. (2017). An Extra-Uterine System to Physiologically Support the Extreme Premature Lamb. *Nature Communications, 8*, 15112.

Pence, G. (2006). *What's So Good about Natural Motherhood?* In S. Gelfand & J. R. Shook (Eds.), *Ectogenesis: Artificial Womb Technology and the Future of Human Reproduction*. Rodopi.

Pfeffer, N., & Woollett, A. (1983). *The Experience of Infertility*. Virago.

Räsänen, J. (2017). Ectogenesis, Abortion and a Right to the Death of the Fetus. *Bioethics, 31*(9), 697–702.

Raymond, J. (1987). Preface. In G. Corea et al. (Eds.), *Man-Made Woman: How the New Reproductive Technologies Affect Women* (pp. 9–13). University of Indiana Press.

Rich, A. (1977). *Of Woman Born: Motherhood as Experience and Institution*. Virago.

Romanis, E. C. (2018). Artificial Womb Technology and the Frontiers of Human Reproduction: Conceptual Differences and Potential Implications. *Journal of Medical Ethics, 44*, 751–755.

Rowland, R. (1987, Spring). Technology and Motherhood: Reproductive Choice Reconsidered. *Signs, 12*(3), 512–528.

Satz, D. (1992). Markets in Women's Reproductive Labor. *Philosophy and Public Affairs, 2*(2), 107–131.

Simonstein, F., & Mashiach-Eizenberg, M. (2009). The Artificial Womb: A Pilot Study Considering People's Views on the Artificial Womb and Ectogenesis in Israel. *Cambridge Quarterly of Healthcare Ethics, 18*(1), 87–94.

Singer, P. (1994). *Rethinking Life and Death. The Collapse of Our Traditional Ethics*. Text Publishing.
Singer, P., & Wells, D. (1984). *The Reproduction Revolution: New Ways of Making Babies*. Oxford University Press.
Singer, P., & Wells, D. (2006). In S. Gelfand & J. R. Shook (Eds.), *Ectogenesis. In Ectogenesis: Artificial Womb Technology and the Future of Human Reproduction* (pp. 9–26). Rodopi.
Smajdor, A. (2007). The Moral Imperative for Ectogenesis. *Cambridge Quarterly of Healthcare Ethic, 16*, 336–345.
Smajdor, A. (2012). In Defense of Ectogenesis. *Cambridge Quarterly of Healthcare Ethics, 21*(1), 90–103.
Smajdor, A. (2016). Ectogenesis. In H. ten Have (Ed.), *Encyclopedia of Global Bioethics*. Springer.
Steiger, E. (2010). Not of Woman Born: How Ectogenesis Will Change the Way We View Viability, Birth, and the Status of the Unborn. *Journal of Law and Health, 23*, 143–171.
Stone, A. (2011). *Feminism Psychoanalysis and Maternal Subjectivity*. Routledge.
Takala, T. (2009). Human Before Sex? Ectogenesis as a Way to Equality. In F. Simonstein (Ed.), *Reprogen-Ethics and the Future of Gender* (pp. 187–195). Springer.
Tonti-Filippini, N. (2003). The Embryo Rescue Debate: Impregnating Women, Ectogenesis, and Restoration from Suspended Animation. *The National Catholic Bioethics Quarterly, 3*(1), 111–137.
Usuda, H., Watanabe, S., Saito, M., Sato, S., Musk, G., et al. (2019). Successful Use of an Artificial Placenta to Support Extremely Preterm Ovine Fetuses at the Border of Viability. *American Journal of Obstetrics and Gynecology, 221*(1), 69.e1–69.e17.
Walters, W. (1982). Cloning, Ectogenesis, and Hybrids: Things to Come? In W. Walters & P. Singer (Eds.), *Test-Tube Babies: A Guide to Moral Questions, Present Techniques and Future Possibilities*. Oxford University Press.
Warren, M. A. (1986). Making Babies: The New Science and Ethics of Conception by Peter Singer; Deane Wells. *Ethics, 97*(1), 288–289.
Warren, M. A. (1989). The Moral Significance of Birth. *Hypatia, 4*, 46–65.
Wishart, B. (1982). Motherhood within Patriarchy: A Lesbian Feminist Perspective. *Third Women and Labour Conference Papers, 1*, 23–31.
Wolfe, A. D. (2008). Wrongful Selection: Assisted Reproductive Technologies, Intentional Diminishment, and the Procreative Right. *Thomas M. Cooley Law Review, 25*(3), 475–502.

Missed Opportunities: Feminist Grounds for Regulating Transnational Surrogacy, in the Anthropocene

Dana S. Belu

This chapter uses key aspects of liberal feminism and feminist phenomenology to advocate for the international regulation of transnational commercial surrogacy, the most popular of all ARTs (assisted reproductive technologies.) Based on my feminist inquiry into commercial surrogate subjectivity, I argue for the need of international regulations on environmental grounds. Commercial surrogacy is a technological practice that compounds harms to the environment during a time of escalating human-caused climate change. If left unregulated, it is an environmentally unsustainable technological practice.

At the time of this writing both India and Thailand, top destinations for reproductive tourism, have already banned foreign commercial surrogacy. But a ban is not a regulation. Whereas feminist inspired international regulations aim for sustainability (by balancing the rights of the surrogates with the rights of future generations on this earth), bans ultimately disregard these concerns. Bans spawn underground markets in the countries of the

D. S. Belu (✉)
California State University at Dominguez Hills, Carson, CA, USA
e-mail: dbelu@csudh.edu

E. Terrone, V. Tripodi (eds.), *Being and Value in Technology*,
https://doi.org/10.1007/978-3-030-88793-3_6

ban and produce new unregulated markets in other developing countries. Furthermore, they leave the surrogates with less money but confronting the same risk factors (Choudury, 2016, pp. 13–14). Bans in the U.K. and Australia,[1] for example, have pushed the practice oversees, to India and Thailand. The current bans in India and Thailand are seeing markets flourish in the Ukraine, Mexico and Nigeria. Bans do not protect the environment or the surrogates, while international regulations aim to do both.

Back in 1976 philosopher Hans Jonas observed that "novel technological powers" in advanced industrialized societies are far surpassing these societies' abilities to make sense of the technologies' moral and environmental impact on the public (Winston, 2014, p. 117). Jonas was right. The gap between technological advances and the ability to make moral sense of their consequences widened with the first successful use of IVF (*in vitro fertilization*) in humans in 1978, the birth of the first test tube baby, Louise Brown. It widened again in 1985 when IVF produced the first surrogate birth, Baby M. A legal battle ensued when the baby's commissioning parents, the Sterns, and her genetic surrogate mother, Mary Beth Whitehead, went to court over custody. By the end of the pregnancy, Mary Beth felt such a strong bond with the baby that she found it impossible to relinquish her, as she had contracted to do many months earlier. In the same year, traditional surrogacy (that uses the surrogate's egg) expanded into gestational surrogacy (that does not use the surrogate's egg) and the first child was birthed by a woman who was not genetically related to the child. That first gestational child and surrogate remain anonymous to this day, yet they helped bring into view a new medical horizon, the age of technologically outsourced reproduction and with it, the reconfiguration of motherhood. A quiet revolution had taken place in the way that humans come to life, or rather, are made to come to life.

Today transnational, commercial surrogacy outsources pregnancy to all corners of the globe. It is a multi-billion-dollar ART industry helping to produce millions of human beings that would not exist otherwise and whose births have unprecedented consequences, including environmental ones. The gateway medical technology to this lucrative business is *in vitro fertilization* (IVF).

---

[1] P. Gerber (2015) "A Human Rights Response to Commercial Surrogacy". https://www.youtube.com/watch?v=lIVRHwZICr0

1 Liberal Feminism and Transnational Commercial Surrogacy

Motherhood was born but the mother was not in the room. (Sophie Lewis)

As the gateway procedure to numerous assisted reproductive technologies (ARTs), including gestational surrogacy, IVF has eschewed regulation. Even most feminist bioethicists (Thompson, 2005, p. 57) do not advocate regulation of IVF- and high-tech IVF-related practices such as commercial surrogacy. They prefer to let the practices regulate themselves, as they putatively do in a free global market. Gestational surrogates, the argument goes, should be allowed the freedom to make their own reproductive choices (Shalev, 1989; Fabre, 2006; Spar, 2006; Parks, 2010; Kirby, 2014) and to engage in what is sometimes euphemistically called "collaborative parenting," even though the vast majority of surrogates do not parent and are not even allowed to meet their offspring.

Since surrogates are not coerced to enter into this contractual relationship, their entry is seen as a sign of individual autonomy that is widely supported by liberal feminists (McElroy, 2002; Gupta, 2006; Spar, 2006; Parks, 2010; Bailey, 2011). From their point of view, regulating surrogacy could seem to violate the personal autonomy of the surrogate if the regulation restricts the surrogate's reproductive freedom, her right to control her body, including the "right to sell her body parts and to gain financially from performing a reproductive service" (Fabre, 2020, p. 2; Spar, 2006; Parks, 2010). By limiting the surrogate's right to sell the use of her reproductive body part, that is, her uterus, it denies her full ownership of her body. Thus, it violates the liberal, Cartesian inspired, albeit controversial,[2] Lockean ontology that an individual is first and foremost a rational subject who *has* a body, and who has a property right in her body (Locke, Treatise II, ch. 5, #27). According to this ontology, the body belongs to "me" but is not identified with me. "I" am separable from my body. Based on this view, "being the owner of this body, a person has the right to do with her body and its parts whatever she wishes,

[2] See Donna Dickenson's *Property Rights in the Body: Feminist Perspectives*, pp. 65–87. Cambridge University Press, 2007.

as long as it does not violate the moral sovereignty of other people (their liberty and integrity)" (Soniewitka, 2020, p. 55). International regulation could seem to impinge on this right if it restricts a surrogate's access to use her reproductive body to earn a living and if it impinges on the commissioning parents' attempts to "secure the good life for themselves" (Fabre, 2020, p. 2) by having a family. Since making a livelihood and achieving parenthood are considered to be central aspects of living a good life, regulations that deny or severely limit an otherwise available livelihood to the surrogate and that curtail a decent chance at parenthood to the commissioning parents count as violations of their interests, their pursuit of the good life (Fabre, 2020, p. 2).

Some liberal feminists[3] are concerned not just that women are free to make individual choices but that they also have the necessary "enabling" conditions to make such choices. This procedural consideration challenges the oversimplified view that the central assumption of all liberal thought is that the individual is prior to society and that all individuals are equal and have equal rights (Oliver, 1989, p. 100). Liberal feminists who advocate a "procedural account" of personal autonomy are concerned that women may suffer from "autonomy deficits"—that is, social and psychological conditions that obstruct free choice. On this account, women cannot be "free and dignified … makers of choices" (Nussbaum, 1999, p. 46) if they are constrained by violence or the threat of violence, stereotyping and discrimination, deprived of vital material goods and are culturally suppressed (Cudd, 2006, pp. 85–118). If we apply these constraints to the typical case of a commercial surrogate in a developing country, such as India (where reproductive tourism was booming until the recent ban), then we can see that a surrogate's choice is not free from material need and cultural oppression. The majority of surrogates live at the poverty line or below. Furthermore, since many surrogates have only a middle-school education, sometimes less than that, and suffer from social expectations to perform motherhood and obey their husbands, the choice cannot be seen to be unconstrained by stereotyping and

[3] Also known as procedural feminists.

discrimination.[4] In fact, the 2019 ban on transnational gestational surrogacy in India signals a victory for those who argue that the practice exploits the surrogates and perpetuates gender inequalities.

The ban, known as the *2019 Surrogacy Regulation Bill*, completely outlaws commercial surrogacy and imposes strict limitations on altruistic surrogacy. It prohibits foreigners, single people and people in same sex couples from becoming commissioning parents. The ban seeks to abolish commercial surrogacy because it is an exploitative practice, an act of "baby outsourcing." Even Indian liberal feminists who criticize the ban on the grounds that it "strips women of autonomy over our own bodies and, in the process, violates our basic civic and fundamental rights" (Kumar, 2019; Lewis, 2019) concede that many surrogates were subjected to "autonomy deficit" conditions. Some of these conditions include "malpractices [being] often underpaid, blackmailed, or extorted, and commonly lacking legal protections and adequate health insurance. Many women become surrogates without having properly verified their fitness to carry a healthy baby to term." However, these liberal feminists believe that even these dire conditions "do not warrant a complete ban". For example, Karuna Nundy, liberal feminist Indian lawyer and advocate on India's Supreme Court, supports the procedural account while also criticizing the ban for paternalistically "telling women what to do" and failing to "respect a woman's right to free choice."[5] In her view a ban will simply drive commercial surrogacy underground where surrogates, especially poor surrogates, will be further exploited, and have their health endangered. Ultimately, according to Nundy, "a poor woman wants her range of options to be expanded, not further limited" (Ibid). While women should have more economic options available to them, she insists that commercial surrogacy remain one such viable option and that the practice ought to be better regulated.

Procedural accounts of autonomy correctly point out that the threat of violence, patriarchy, poverty and cultural oppression are real obstacles to women's autonomy. These are sedimented social conditions that result

---

[4] See work by K. Vora, A. Pande, A. Bailey, S. Lewis and SAMA on actual and potential harms accrued to gestational surrogates. This work lies beyond the scope of my essay.

[5] See Al Jazeera's *Inside Story* with Hazem Sika, August 25, 2016, "Why is India Banning Commercial Surrogacy?". https://www.youtube.com/watch?v=IpuVaQ-r1SY

from reified sexist relationships that naturalize gross levels of gender inequality. Nandita Rao, Indian lawyer and civil rights activist, speaks in support of the ban and in an attempt to raise consciousness about surrogate exploitation, what she calls the "slavery" of Indian surrogates: "Surrogates are being forced by their husbands to do this work without any insurance from the clinic or the state that they will get paid if they get sick or lose the baby. The power differential between surrogates and the commissioning parents is huge, like 90% to 10%. Is this really informed consent? If you are saying she is making a free choice to be a surrogate you are making a joke of yourself and the idea of free choice. If [surrogates] are in such a desperate situation that they have to sell their womb then the state has to step in and give them a human life.... It is the job of the state to protect the weak."[6]

Raising consciousness about the constraints of choice is significant since it brings into view the socio-economic-political deficits in the lives of many surrogates. If Rao and other like-minded critics are right, then at best surrogates make *conscious choices* but not *free choices*. That is, they do not make choices that are unconstrained by the will and power of others. Surprisingly, neither a liberal feminist critique of the ban nor a radical feminist endorsement of the ban, such as Rao's, considers the environmental unsustainability of this practice and the need for including environmental protections in international regulations. These regulations ought to protect the environment, in addition to protecting the surrogates, and they ought to be recognized by all participating states. Fortunately, procedural concerns that draw attention to the background conditions of choice help bring into view the greatest *dis-abling* condition for free choice, a heavily polluted, unlivable environment, to which transnational commercial surrogacy is still steadily contributing.

Unlike the more obvious harms done to surrogates, the harm done to the environment conceals itself. Heavy carbon footprints resulting from transnational flights,[7] additional production of medical equipment, the routine births of twins or triplets, are difficult to see and to track. When

---

[6] Ibid.

[7] Geerts, Robert-Jan (2018), "Climate Change and the Philosophy of Technology." In *Spaces for the Future: A Companion to the Philosophy of Technology*, eds. Joseph Pitt and Ashley Shew. Routledge Press, p. 295.

the impact of these externalities is actually tracked by environmental scientists and statisticians, they are treated as acceptable externalities by the law, the mainstream media and population. This is evidenced by the fact that neither the clinics, the commissioning parents, nor the surrogates are held responsible. Some of the participants may even be ignorant of the consequences of their actions and the law finds their ignorance of no punishable consequence. But the consequences are real and they adversely impact everyone. While it is difficult, if not impossible, to point to specific individuals who will be most directly affected by these increased emissions, polluted air and water, inundations and fires present an immediate and "direct harm" to almost anyone, (Cudd, 2004, p. 54) human and non-human others.[8] Unfortunately, women and children living in the developing world, including India and other countries in the global South East, are especially vulnerable to the harmful effects of climate change. Thus, fighting climate change is not a gender-neutral concern but lies at the heart of feminism.

Being unable to: breathe clean air, go outdoors, travel freely, enjoy nature, live a life free of asthma, stay in your house without threat of floods or fires, are all recent examples of severe autonomy *dis-abling* conditions as a result of anthropogenic climate change. None of us, on our own, is free to choose a way around these threats to our shared human habitat and personal safety. Striving toward sustainability requires that people learn to meet the needs of the present without increasing carbon emissions and compromising the ability of new generations to meet their survival needs in the future. This must be a global effort in order to succeed. This effort is derailed by the use of vast resources to bring children into a world that they environmentally undo before they even come to life. As one environmental activist put it, in the current environmental crisis, "any diversion of material and energy to grow more humans and their 'furniture' ... [and] the increased production/consumption for humans adds to the pollution load on natural ecosystems."[9]

[8] This includes non-human others such as various forms of wildlife, frogs and fish, whose fertility gets disrupted by the dumping of IVF fertility drugs into rivers and lakes. See Legal (2015, p. 82).

[9] Cited in Portnoy's *Sustainability*, 26.

All enabling, background conditions of personal autonomy are indebted to what I call the meta-condition, the background condition of having a clean, life-supporting environment. When the consequences of some individual freedoms destroy this background condition, international regulations must step in. While procedural feminism does not directly focus on the climate as the meta-condition for all social conditions upon which personal choices, including women's choices, depend, its focus on enabling social conditions helps to bring this all-important meta-condition into view.

Procedural feminism is grounded in a commitment to a woman's personal autonomy and a respect for other women's choices. It steers clear of promoting illiberal choices that impose prescriptions or make substantive demands about how women should live their lives. From the procedural perspective, the recent ban in India is prescriptive since it severely limits women's choices but does not improve social conditions that would open up other, life-affirming choices for these groups of women. It deprives surrogates of needed income (Saravanan, 2018, p. 131), drives the market underground where the surrogates are more vulnerable than ever, and exports commercial surrogacy to other developing nations. So, the ban does not address the problems it purports to address, ending surrogate exploitation and "baby outsourcing." It merely pushes the problem around, leaving "the exploitation of poor women and lack of protection from unfair domestic surrogacy contracts unaddressed" (Choudury, 2016, p. 13).

So, from a procedural liberal perspective the ban cannot be justified because it frustrates the exercise of women's individual autonomy and does not provide new or better enabling conditions for women to exercise said autonomy. The procedural feminist recasting of Mill's liberal principle states that "The only justification the state has for coercing individuals to act contrary to their preferences is to prevent harm, including the oppression of others" (Cudd, 2004, p. 54). This could be seen to endorse the ban if and only if it also provides new economic opportunities for the women it puts out of work. However, it does not provide any such opportunities. Enabling conditions of choice such as relative material stability, cultural freedom, education and free speech strengthen personal autonomy even as these exercises in autonomy in turn strengthen the enabling conditions.

Unfortunately, neither the current ban nor its critics mention the environmental harm posed by a poorly regulated system of transnational surrogacy. Expanding the procedural requirements discussed above means acknowledging that a clean environment is an equally important, if not the most important long-term "autonomy enabling condition" for making choices, especially in the midst of a climate crisis. A feminist phenomenology that brings out the overlooked technologically embodied dimension of commercial surrogacy will make this clearer.

2 Feminist Phenomenology and Transnational Commercial Surrogacy

Gestation is not merely a biological or physiological process, but instead a complex process of affective co-constitution. (Luna Dolezal)

2.1 Embodiment

As I've shown above, liberal feminism focuses on the individual rights of the surrogate and on making the necessary social conditions available so that women can exercise their rights. However, it ignores the embodied experience of pregnancy. This is a missed opportunity that feminist phenomenology seizes upon. A feminist phenomenological ontology is needed if we are to gain an understanding into the constitution of the surrogate's pregnant subjectivity and its technological mediation that in turn exposes commercial surrogacy as a source of anthropogenic harm.

Liberal feminist and feminist ethnographic literature on commercial surrogacy frame the gestational surrogate fairly homogeneously, either as a rights bearing, rational subject or as an exploited object.[10] The language used is that of body parts, individual rights and the freedom of the

[10] Alison Bailey's groundbreaking 2011 article "Reconceiving Surrogacy: Toward a Reproductive Justice Account of Indian Surrogacy" brings out the potential moral absence in feminist ethnography and the colonialist imposition inherent in normative, western feminisms in commercial sur-

Cartesian mind to engage in acts of reproductive freedom. These acts allow the surrogate to use her body as property for financial gain. As a rational subject she is seen as a savvy entrepreneur who makes the most of her scant economic options in a society ruled by patriarchy and sexist exploitation. As an exploited object she is seen to lack the necessary social, economic and political conditions to make an informed decision. Unfortunately, both accounts miss the embodied experience of pregnancy. Drawing on the phenomenology of Maurice Merleau-Ponty, feminist phenomenology investigates this overlooked dimension as the *intercorporeality* that structures the maternal-fetal dyad. Access to this dyad reminds all of us, despite our inevitable forgetfulness, "that our first experiences are ones of inseparableness from our mothers *in utero*" (Welsh, 2018, p. 296).

In "Phenomenology and Intercorporeality in Commercial Surrogacy", Luna Dolezal explains this *in utero* "inseparableness" as "a tacit communication involving an embodied entanglement between self and other that constitutes the possibility for individual subjective experiences" (Dolezal & Petherbridge, 2017, p. 320). To this intercorporeality or pre-reflective "intertwining with the other" (Dolezal & Petherbridge, 2017, p. 321) belongs the body schema, or "the subject's noncognitive awareness of its position, orientation and movement" (Dolezal & Petherbridge, 2017, p. 322). One's body schema is fundamental to motor intentionality and it is flexible, "renewed through the sedimentation of tacit skills and techniques that makes repeatable action possible" (Dolezal & Petherbridge, 2017, p. 322). The body schema is not a body image, since it is not a mental representation of any sort. Rather, it refers to a pre-reflective mode of relating to the world through movements and attitudes that get deposited and integrated, better yet, absorbed into a changing and flexible, yet habitual, body or "register."

Drawing on Jane Lymer's phenomenology of gestation in "Merleau-Ponty and the Maternal-Foetal Relation", Dolezal underscores that the commercial surrogate's maternal body schema is necessary for the incorporation of the fetus and ultimately for the formation of the body schema

rogacy. Also see J. Kirby's "Transnational Gestational Surrogacy: Does it Have to be Exploitative?" *The American Journal of Bioethics*, 14(5): 24–32, 2014.

of the child (Dolezal & Petherbridge, 2017, pp. 323–324). Lymer argues that newborn infants possess a "primitive body schema" that they learn *in utero.*[11] This means that the fetus acquires this body schema from intertwining with the surrogate mother. This "affectively structured, embodied relation guides the fetus, and possibly then the child, through the early stages of subjectivity development… It constitutes the ability to form attachment bonds outside the womb" (Dolezal & Petherbridge, 2017, pp. 324–325). Lymer's phenomenology of gestation goes further and enlists empirical evidence from infant neurology in support of her thesis that the baby's body schema begins to form in the first trimester rather than post-partum, as previously thought by Merleau-Ponty and others. She shows that through a continuous renegotiation of her habituated movements, including adjusting to her growing belly, the lilting rhythm and slowing pace of her walk, "the maternal body provides for the fetus a primal interaction that molds and forms the fetal body schema as a kind of *imprinting*" (Lymer, 2011, p. 9, my emphasis). This body schema accounts for the neonate's organized responses to her new environment through acts of "invisible imitation" (Ibid). The significance of Lymer's work is to underscore the depth of the surrogate's embodied contribution to the newborn. One central aspect of this embodied contribution, I contend, is the voice.

In *The Visible and Invisible*, Merleau-Ponty identifies one's own/my voice as a "movement" that we hear *with* the "throat" and that is "bound to the mass of my own life as is the voice of no one else" (Merleau-Ponty, 1968, p. 144). This synesthetic experience of myself as embodied mingles the objects of my sense perceptions, in this case revealing the tactility of hearing. As a "sonorous" being, my voice traverses my body, is my body, and like all my other bodily rhythms is necessarily felt, absorbed by the fetus who hears/feels the mother's movement with her whole body. The

[11] Dolezal draws on Lymer's (2011) feminist phenomenological ontology according to which "the maternal body schema forms the basis of and for the foetal body schema and subsequent foetal development. Together, the manner in which the maternal and foetal body schemas merge and then diverge will form a communication that is born through situated, gestational embodied negotiations. This relationship is affectively structured through the negotiated movements themselves. Thus, by the time of our birth we have already, within our habituated repertoires, a way of moving and interrelating that may well set the foundations for affective intersubjective relations post-partum." *Parrhesia*, 142.

mother, then, by definition *is* the one whom the fetus hears, feels and mimics. Luce Irigaray underscores this point in her reading of Merleau-Ponty foregrounding the primacy of a pre-discursive, tactile, palpating communication between the mother and the fetus: "In utero, I see nothing (except darkness?), but I hear. Music comes before meaning. A sort of preliminary to meaning, coming after warmth, moisture, softness, kinesthesia. Do I hear first of all? After touch. But I cannot hear without touching: nor see, moreover. I hear, and what I hear is sexually differentiated. Voice is differentiated" (Irigaray, 1993, p. 168). Tactile and visual communication implies differentiation, and differentiation, the repeated picking out of a "this" rather than a "that" defines all embodied relationships, beginning *in utero*. Prior to any concrete, expressive content, even *prior* to reciprocity, the fetus[12] "picks out" *the materiality* of the surrogate's voice,[13] her timbre and rhythm, all of which provide a first constancy and connection, an imprinting.

By the second trimester, the connection between the surrogate mother and the fetus, their "tactile sociality" becomes visibly reciprocal. The baby kicks and the mother strokes her belly, the mother speaks, sings or cries and the baby responds. By the time the baby is born, a pre-reflective, affective bond is in place. The baby's responses help to define the ontological maternal-fetal intercorporeality, that pre-reflective, embodied "entanglement between self and other" that subtends reflective life and forms the core of each fetus's "*prenatal sojourn*" (Irigaray, 1993, p. 165).

The phenomenology of gestation explains the attachment that Indian gestational surrogates report in recent ethnographic studies. Surrogates speak of feeling an intense bond with the babies and that the babies respond to their voices, are soothed by them. They do not respond to the voices of the commissioning parents (Saravanan, 2018, p. 138). Many

[12] According to Barbara Duden's haptic phenomenology, the fetus *as* a being that exists separate from the mother is "an object of our time" unknown to previous historical epochs. According to these earlier historical accounts, the "fetus" was an aspect of the pregnant woman's body. See "Disembodied Health" in *Thinkery*, 2000 and "The Fetus as an Object of our Time" in *Anthropology and Aesthetics*, No. 25 (Spring, 1994), pp. 132–135. I am thankful to the blind reviewer who recommended Duden's work.

[13] For a phenomenological investigation into the material voice and its difference from the expressive voice, see Linda Fisher's "Feminist Phenomenological Voice" in *Continental Review* (2010) 43:83–95.

surrogates speak of themselves as surrogate mothers (Pande, 2014) and report a deep sense of longing and loss when the babies are whisked away. Some surrogates simply want to "have the children back" (Saravanan, 2018, p. 118), saying that the experience of separating from the newborns felt traumatic, like "giving away a part of my body and my soul." One referred to her surrogate newborn as "the womb-brother of my other son" (Saravanan, 2018, p. 124). Yet their suffering is minimized. According to the clinic and to the intended parents, the surrogates are not "giving anything up" since the babies were never theirs to begin with. The pain of separation is seen as unjustified and thus easily dismissed along with the bond.

But experiences of the maternal-fetal bond are so pervasive that women in the global North West pay to experience it. Older women in the US who can no longer use their own eggs to have a baby, choose gestational pregnancy with an embryo formed from the ovum of a younger woman. The older woman acts as the gestational "surrogate" who will also be the social mother. In pregnancy she sees herself and is seen by others as the mother. In fact, this is one of the dual purposes of this non-genetic pregnancy, to experience the material intimacy of pregnancy and to gain the social authority denied by the genetic deficit. More recently women born without a uterus or those with hysteroctomies are opting for uterine transplants for the purpose of gestating their own children. They join an expanding group of women who desire to experience the materiality and intimacy of pregnancy,[14] seeing it as a form of mothering.

The demand for gestating one's own non-genetic baby gives the lie to the normative view that assesses surrogacy *primarily* as a rights-based, instrumental contract. Contrary to this instrumental view, feminist phenomenology points to a primary, affective experience between the mother and the fetus, an early bond that cannot be denied, though it can be differently valued. The desire for this bond reveals that pregnancy is seen as a form of mothering. Unfortunately, the double standard in gestational surrogacy makes it look as a form of mothering when wealthy, white women of the North West do it and as just an instrumental transaction

[14] https://www.pennmedicine.org/news/internal-newsletters/system-news/2019/may19/uterus-transplants-a-new-door-opens

involving movable body parts when brown women in the global South East do it (or when poor women do it). One could object that the bond is conditioned by the expectation of raising the baby, knowing it will be a part of one's family. Perhaps this is a contributing factor but it is not the only one, since as we have seen above, feminist ethnographers report that the bond is also felt intensely by surrogates who know that they will not raise the child and who are psychologically trained to dissociate from the fetus. While feminist phenomenology and ethnography account for the bond, they miss the opportunity to describe its technological mediation.

2.2 Technology

Since technology is the indispensable medium of commercial surrogacy and since the surrogate fetus is the exclusive product of technological interventions, a feminist phenomenology of surrogacy ought to account for this mediation of the bond. Uncovering the technological mediation brings into view anthropogenic environmental harm and the need for environmental regulation.

In "The Question Concerning Technology", Martin Heidegger identifies the ontological characteristic (*Grundzug*) of the technological age as a "mode of revealing" (*aletheuein*, *Entbergen*) (Heidegger, 1977, p. 14). More specifically, he describes this revealing as a "challenging-forth" an attitude that reduces nature and things to a heap of fungible raw materials or "standing reserve [that] no longer stand over against us as object (*Gegenstand*)" (Heidegger, 1977, p. 17). In earlier work,[15] I introduced the term *reproductive enframing* to refer to the challenging forth of women's bodies, their decontextualization and reduction to their reproductive parts and potential. *Reproductive enframing* sums up the manipulation of this potential by describing a fragmented approach to conception and gestation; one that frames the uterus as a collection of discrete and movable reproductive parts: ovaries, follicles, eggs, fallopian tubes and hormones. These parts are managed as stock, potential reproductive energy that is challenged forth in the Petrie dish. There the "energy" of the sperm

[15] See my *Heidegger, Reproductive Technology and the Motherless Age* (Palgrave, 2017) pp. 24–51.

and the egg is "unlocked" to achieve fertilization and then "stored-up" until the embryo is implanted in the uterus of the surrogate, who is standing reserve, "on call for a further ordering" (Heidegger, 1977, pp. 16–17). *Qua* standing reserve, each surrogate shows up as interchangeable with other women in a transnational fertility network that operates without her *qua* individual person. The gestational surrogate is the paradigmatic human resource of the technological age. She stands outside of the subject-object order.

Since she is being paid for services that she agrees to render, she cannot be considered an object (or a slave). By definition, slaves are not compensated for their labor and are not consulted about the terms and conditions of their labor, while surrogates are consulted. The surrogate, however, is not quite a subject either since a host of "autonomy deficit" conditions, such as poverty, lack of education and oppressive patriarchal constraints define her choice. Of special note is the restriction in her ability to understand, and thus freely consent to, the terms of the contract. In India, contracts are often written in English and are then translated for the surrogate who is not "independently represented." This presents a "procedural problem" where the "language barrier can lead to a one-sided distribution of risks and liabilities" (Choudury, 2016, p. 8) and worse, an inadequate understanding of the "assumed" medical risks.

Moreover, choosing surrogacy does not necessarily increase the woman's autonomy since she is making another choice related to a patriarchally sanctioned motherhood function. She is not engaging in a "genuinely alternative life option."[16] In fact, she agrees to surrender many of her freedoms, especially her *embodied* freedom(s) in order to be a surrogate. This surrender amounts to a series of acts of self-objectification that further undermine her subject status and underscore her resource status. For example, her fungibility is revealed when she "signs off all rights on medical intervention" (Saravanan, 2018, p. 113), thereby giving the clinic legal, not merely medical, access to her body. In practice, this means that she agrees to move into a surrogate hostel and to undergo complete surveillance (Bailey, 2011, p. 722). She agrees to forego termi-

[16] Cited in Christine Legals' *An Ecofeminist Analysis of In Vitro Fertilization*, p. 64. Master's Thesis. University of St. Michael's College, Canada, 2015.

nating her pregnancy (Saravanan, 2018, p. 117). She agrees that in "life threatening situations [she] must subordinate her well-being to that of the child" (Saravanan, 2018, p. 122). In these extreme situations, as in other routine ones, the well-being of the child is ensured by the use of medical technologies. Yet, since the child is still *in utero*, any medical access to the child must go through her body. She is the medium for the technological being of the fetus' and its well-being but has no say in these procedures and their impact on her body. Yet, choosing this surrogacy means choosing the intertwining of her body with a multitude of technological procedures and machinery.

From the perspective of a feminist phenomenology of technology omitting this technological mediation, as does Dolezal in her otherwise excellent article, has the unintended consequence of assimilating the structure and experience of commercial pregnancy to traditional pregnancy. While non-IVF-based pregnancies are also mediated by technologies in the form of routine check-ups and tests, including the monitoring of hospital births and the pressure to undergo caesarians, the ontological difference between the two kinds of pregnancy is significant. Since the non-surrogate contributes her own egg, she is not framed as fungible. In addition, her pregnancy does not *decontextualize*[17] her from her lifeworld and *reduce* her to her reproductive potential. Finally, unlike her gestational counterpart, she retains significant aspects of her autonomy during pregnancy and birth.

In addition to undergoing invasive medical procedures, the gestational surrogate is uniquely subjected to psychological messaging that trains her not to see herself as connected to the fetus, that she is not the baby's mother but a "mere vessel" akin to occupied territory.[18] Following the

---

[17] See Andrew Feenberg's for his *primary and secondary instrumentalization* theory (1999, 2010). that concretizes the enframing. It avoids a poorly differentiated, reductive understanding of technology by foregrounding the connection between the three stages of secondary instrumentalization (or the social integration of new technologies) made up of: systematization, mediation and vocation, and the four stages of primary instrumentalization: decontextualization, reduction, autonomization and positioning (the function level.)

[18] See Kalindi Vora's interpretation of commercial surrogacy as an extension of Western colonialism in South East Asia, and a perpetuation of colonialist attitudes, a racial imaginary, that sees open land as available for taking and by analogy, sees women's wombs as potential property. "Medicine, Markets and the Pregnant Body: Indian Commercial Surrogacy and Reproductive Labor in a Transnational Frame" (p. 5).

clinic's script requires a kind of splitting, a depersonalization and *autonomization* of her pregnancy, the interruption of the *reflexivity* of technical action, its impact on the user. It involves psychological techniques of repetition and disassociation that can be traumatic, especially for those who do not internalize the script well and/or suffer post-partum duress. *Autonomizing* the surrogate's psychology is clearly visible when the clinic dismisses the surrogate's post-partum distress and adopts an administrative or "purely functional" relationship to her. This affords the clinic immunity from the consequences of its actions and casts the woman as fungible. Whatever the psychological result may be, the dual impact of the technological machinery and that of the psychological messaging techniques that she is a "mere" container, a conduit, reactivate old stereotypes about women and also frames the surrogate as a resource.

Continuous technical interventions such as excessive watchfulness and monitoring by the medical staff and herself, transform the surrogates' lived body (*Leib*) into an object body (*Körper*),[19] the body that is medically re-presented and given back to her as a set of facts. For instance, the ubiquitous reliance on ultrasound testing to "see" and to "evaluate" the fetus's health, that is, size, heart rate and cognitive development, among others, reinforces the dominance of medical equipment and devalues the surrogate's subjective experience. She is not consulted and thus an objective body is foisted onto her lived,[20] phenomenological body even as the technicization of the latter is paradoxically ignored, passed off as natural. Yet the pre-reflective and continuous renegotiation of the surrogate's body schema in pregnancy involves subtle ways of coping as the lived body accommodates the technological gestation. I wonder if the continu-

[19] See Edmund Husserl's phenomenological distinction between the lived body *(Leib)* and the physical or objective body *(Körper)*. See Husserl's *Ideas II*, especially pp. 150, 155, 159, 166.

[20] For a heterosomatics and a historico-phenomenological account of pregnancy as *haptic*, *humoral* and defined by a "quickening" unlike the *visual* and organ centric account of pregnancy defined by a "due date", see the work of Barbara Duden. Duden argues that medical categories such as *the fetus*, *life* and *the womb* would be unintelligible to pre-twentieth-century birthing practices when a woman's narrative was part of her pregnancy. "What each woman brought to her doctor was not a body to be inspected and palpated but a story that is enfleshed." (Duden 1993, p. 135) Back then "before a child comes to light it is a *nondum*, a "not yet." (Duden, 2000, p. 10) A being of hope." It is not a fetus. By the late twentieth century, "the technician foists on the [pregnant] patient a configuration of parameters, an iatrogenic (medicine made) construct that the patient ...ascribes to [herself.]" (Duden, 2000, p. 58)

ous technological monitoring and psychological messaging, restrictions in the surrogate's pre-reflective movements, have an adverse impact on the renegotiation of her maternal body schema and on the "imprinting" of the fetus's body schema?

When intercorporeal accounts of pregnancy, such as Dolezal's, ignore the technical mediations, the maternal-fetal relationship is naively described as a seamless, natural whole. Thus, the technicized, lived body of the surrogate is merely bypassed, supplanted by a pre-technical pregnant body that is not represented as an object-body but is not fully accounted for in its lived dimension either. The visibility of the surrogate's technicized pregnancy pulls into view the technological network that subtends the transnational surrogacy industry, an unregulated industry that silently pollutes the planet.

The surrogate's world is a technological interlacing of virtual communications, a network of commissioning parents, doctors and clinic staff, air travel, assessment of medical risks and success rates, the signing of contracts and the anticipation of payments. However, what enables the surrogate network to come into being in the first place, is not merely the successful empirical coordination of various technologies but a shared seeing, a pre-reflective framing of women's reproductive ability and of the embryos as fungible, bioavailable resources. The transaction between the "parents", the clinic and the surrogates aims at making the most of their options and in late capitalism this appears to make good sense. What is rendered invisible in this transaction is the use of unregulated technologies that increase anthropogenic carbon emissions during a global, environmental crisis.

3 Regulation

Seen through the lenses of liberal and phenomenological feminisms, a ban on transnational surrogacy is a missed opportunity to understand the technological dimension of commercial surrogacy, to care for the surrogate and for the planet. International regulations informed by feminist phenomenological and procedural accounts can seize upon this missed opportunity as they reveal the technologically embodied dimension of transnational surrogacy and the significance of a clean environment as a fundamental autonomy enabling condition.

In line with the phenomenological account, international regulations ought to recognize the surrogates' roles in pregnancy and after the birth of the baby. In view of the depth and toll involved in each surrogate experience, both contract law and state law ought to recognize the maternal work (mothering) done by surrogates. Developing international laws for protecting surrogates and the newborns ought to involve consultation with the surrogates, on a voluntary basis, and to allow for the possibility of maintaining some kind of kin relationships with the children they brought to life.[21] This provision could protect against the reckless institutional "dumping" of emotional and psychological externalities onto the surrogates and possibly onto the newborns themselves.[22] In addition, international regulations ought to track and reduce environmental externalities (unintended consequences) associated with commercial surrogacy.

In the first section of the paper, I mentioned a triple environmental externality made up of multiple unintended births, air travel and the production of additional medical equipment, all major sources of pollution. For instance, the possibility of twins or triplets as a result of having multiple embryos implanted is roughly 50%.[23] When these newborns travel back to the first world country of the commissioning parents, they begin building the heaviest carbon footprint in the world, and so do their parents, since the carbon footprints of the parents multiply in relation to their biological children. Since the *carbon footprint* of each parent in first world countries is exceedingly high due to over consumption of fossil fuel, meat, dairy and heavy use of high-tech devices, their carbon footprint in relation to their children will also be exceedingly high. The

---

[21] Such kin relations would have to consider the needs, desires and well-being of the children, the surrogates and be worked out in accordance with state and local laws.

[22] Date on the experiences of surrogate children is still rather scant but beginning to be available in recent years, notably from grown surrogate children growing up in Australia.

[23] Due to poor data tracking, an accurate account of multiples is not available. In some cases, when multiples are not desired, the doctors perform a "medical reduction" removing one or more of the embryos. Some parents, as in the case of baby Gammy, do not want a twin if this one has a genetic abnormality. Some commissioning parents simply abandon developmentally challenged newborns on the streets of India. (Saravanan 2018, p. 171).

carbon legacies of all parents *grow exponentially* with the birth of each child.[24] In fact, "under the current emissions scenario, an extra child born to a woman in the United States ultimately increases her carbon legacy by an amount of 9441 metric tons; that is nearly seven times the analogous quantity of a woman in China (1384 tons) …[and 168 times the quantity of a woman in Bangladesh, 56 tons]."[25] Having one fewer child amounts to "a 58 tonnes reduction of carbon dioxide emissions per year compared to a mere 2.4 tonnes yearly reduction from not driving a car."[26] Children living in first world countries are the highest source of carbon emissions on the planet. The more unregulated surrogacies take place, the higher are the chances of multiple births and the more the planetary carbon emissions will rise. While the rich contributions of these newborns to their families and communities can hardly be quantified, a concern with the sustainability of the planet requires that people learn to meet the needs of the present without compromising the ability of new generations to meet their survival needs in the future. This must be a global effort in order to succeed, but it is derailed by the unregulated use of vast resources to bring children into a world that they environmentally undo before they even come to life. Thus, the problem is not one of overpopulation but one of unregulated technological reproduction.

To this belongs the unregulated production, shipping and disposal of a variety of medical equipment necessary for commercial surrogacy,[27] including air travel. The latter is the third worst cause of carbon emissions. Airplanes put tons of carbon-warming greenhouses into the envi-

---

[24] Paul A. Murtaugh and Michael G. Schlax. "Reproduction and the Carbon Legacy of Individuals." *Global Environmental Change* 19 (2009):14. "Our basic premise is that a person is responsible for the carbon emissions of his descendants, weighted by their relatedness to him. For a descendant that is n generations removed from the focal individual, the weight is ð1=2Þ^n. So, for example, a mother and father are each responsible for one half of the emissions of their offspring, and 1/4 of the emissions of their grandchildren".

[25] Seth Wynes and Kimberly A. Nicholas, "The climate mitigation gap," *Environmental Research Letters* 12, No. 7 (2017): 2.

[26] Ibid. 18.

[27] For instance, the growing use of *time lapse embryo imaging* technology helps to pick out viable embryos to increase success rates of IVF and to reduce the birth of multiples. However, this popular technology is likely to increase transnational surrogacy and thus heightens the need for its regulation. For more on this reproductive technology, see Lucy van de Wiel's "The Datafication of Reproduction: Time lapse embryo imaging and the Commercialization of IVF."

ronment each year. This unintended consequence is avoidable by reducing the number of transnational surrogacies and enforcing more environmentally conscious eligibility requirements.

In sum, fewer biological children for some means lower emission levels for all. The freedom for *some* to sell their reproductive services and body parts for money in "the pursuit of the good life" must be balanced by choices that do not worsen the conditions of life for *all*. Hence, the personal freedom to carry an ART baby and to take one home must be weighed against its consequences. Whether or not a surrogate freely chooses to sell her reproductive services, and whether or not she is coerced by her husband, his family or a life of dire poverty, the environmental harm that results from her choice is the same. So, restrictions are needed.

My call for sustainability and environmental justice may appear like an appeal to a utilitarian trade-off, one that merely weighs the needs of future generations against the needs of the present, a calculus that conflicts with my Heideggerian inspired, feminist phenomenological diagnosis of commercial surrogacy as a source of anthropogenic harm. However, this is actually not the case. The move to regulate rather than to ban or to blindly drive commercial surrogacy forward is not about "autonomizing the present, turning it into a calculable array of forces, all for the sake of ordering the survival of new generations."[28] Rather, I conceive of regulation as a form of consciousness raising, a way of getting clearer in our practices, not just in our heads, about the phenomenological inversion at play in commercial surrogacy, namely that the technology is not so much serving the surrogates as the surrogates and the environment are in service to the technologies, even to those who wield them. Anthropogenic harm is a significant by-product of this inversion. This becomes visible once you look at commercial surrogacy phenomenologically and on a continuum with other the use of other reproductive technologies, rather than discretely.

Both the banning of commercial surrogacy and its uncritical promotion are reactionary attitudes that cover up the enframing (Heidegger, 1977, p. 25) by either blindly "pushing it forward" or banning it. We can

[28] My thanks to Duncan Cordy for this critical comment at *The Post Kantian Bioethics Conference*, Purdue University, April 2021.

see that neither approach is working since international markets are expanding yet the working conditions of the surrogates are not improving. A regulatory alternative would build legal protections for surrogates and the environment on feminist foundations by paying attention to the ontological and personal dimensions of surrogacy, including the children, the surrogates, the earth and the air, rather than continue to treat all of these as extractable resources.

References

Bailey, A. (2011). Reconceiving Surrogacy: Toward a Reproductive Justice Account of Indian Surrogacy. *Hypatia, 26*(4), 714–741.

Belu, D. (2017). *Heidegger, Reproductive Technology & The Motherless Age.* Palgrave Macmillan Press.

Choudury, C. (2016). Transnational Commercial Surrogacy: Contracts, Conflicts, and the Prospects of International Legal Regulation. Printed from *Oxford Handbooks Online*, Oxford University Press. https://doi.org/10.1093/oxfordhb/9780199935352.013.38.

Cudd, A. (2004). The Paradox of Liberal Feminism: Preference, Rationality and Oppression. In A. Baehr (Ed.), *Varieties of Liberal Feminism* (pp. 37–61). Rowman & Littlefield.

Cudd, A. (2006). *Analyzing Oppression.* Oxford University Press.

Dickenson, D. (2007). *Property Rights in the Body: Feminist Perspectives.* Cambridge University Press.

Dolezal, L., & Petherbridge, D. (2017). *Body/Self/Other: The Social Ontology of Social Encounters.* SUNY Press.

Duden, B. (1993). *Disembodying Women: Perspectives on Pregnancy and the Unborn.* (L. Hoinacki, Trans.). Harvard University Press.

Fabre, C. (2020). *Surrogacy. International Encyclopedia of Ethics* (H. LaFollette, Ed.). Wiley & Sons, Ltd

Feenberg, A. (1999). *Questioning Technology.* Routledge Press.

Fisher, L. (2010). Feminist Phenomenological Voices. *Continental Review, 43*, 83–95.

Gupta, J. A. (2006). Towards Transnational Feminisms: Some Reflections and Concerns in the Relation to the Globalization of Reproductive Technologies. *European Journal of Women's Studies, 13*(1), 23–38.

Heidegger, M. (1977). The Question Concerning Technology. In *The Question Concerning Technology and Other Essays*. Translated W. Lovitt. Harper & Row Publishers.
Irigaray, L. (1993). The Invisible of the Flesh: A Reading of Merleau-Ponty, The Visible and the Invisible, 'The Intertwining—The Chiasm'. In *An Ethics of Sexual Difference*. Translated C. Burke & G. C. Gill. Cornell University Press.
Jonas, H. (2014). Technology and Responsibility: Reflections on the New Task of Ethics. In M. Winston & R. Edelbach (Eds.), *Society, Ethics, & Technology*. Wardsworth Press.
Kirby, J. (2014). Transnational Gestational Surrogacy: Does it Have to be Exploitative? *The American Journal of Bioethics, 14*(5), 24–32.
Kumar, S. (2019). India's Proposed Ban is an Assault on Women's Rights. *The Wire*. https://thewire.in/law/surrogacy-ban-assault
Lewis, S. (2019). Surrogacy as Feminism: The Philantrocapitalist Framing of Contract Surrogacy. *Frontiers: A Journal of Women's Studies, 40*(1), 1–38.
McElroy, W. (2002). *Liberty for Women: Freedom and Feminism in the Twenty First Century*. Ivan R. Dee & The Independent Institute.
Merleau-Ponty, M. (1968). *The Visible and the Invisible*. (A. Lingis, Trans.). Northwestern University Press.
Murtaugh, P., & Schlax, M. (2009). Reproduction and the Carbon Legacy of Individuals. *Global Environmental Change, 19*, 14–20.
Nussbaum, M. (1999). *Sex and Social Justice*. Oxford University Press.
Parks, J. A. (2010). Care Ethics and the Global Practice of Commercial Surrogacy. *Bioethics, 24*(7), 333–340.
Saravanan, S. (2018). *A Transnational Feminist View of Surrogacy Biomarkets in India*. Springer.
Shalev, C. (1989). *Birth Power: The Case for Surrogacy*. Yale University Press.
Soniewitka, M. 2020. Ethical and Philosophical Problems Arising from Surrogate Motherhood. *Fundamental and legal problems of surrogate motherhood: Global Perspectives*.
Spar, D. (2006). *A Baby Business: How Money, Science and Politics Drive the Commerce of Conception*. Harvard Business School Press.
Thompson, C. (2005). *Making Parents: The Ontological Choreography of Reproductive Technologies*. MIT Press.
van de Wiel, L. (2019). The Datafication of Reproduction: Time Lapse Embryo Imaging and the Commercialization of IVF. *Sociology of Health & Illness, 41*(S1), pp. 193–209. ISSN 0141-9889. https://doi.org/10.1111/1467-9566.12881

Vora, K. (2011). Medicine, Markets and the Pregnant Body: Indian Commercial Surrogacy and Reproductive Labor in a Transnational Frame. In R. Jordan-Young (Ed.), *S&F Online* (Special ed.). Critical Conceptions: Technology, Justice and The Global Reproductive Market.

Welsh, T. (2018). The Order of Life: How Phenomenologies of Pregnancy Revise and Reject Theories of the Subject. In S. LaChance Adams & C. R. Lundquist (Eds.), *Coming to Life: Philosophies of Pregnancy, Childbirth, and Mothering.* Fordham University Press.

Part III

Aesthetics

Computer Art, Technology, and the Medium

Christopher Bartel

1 Introduction

"Art" is a word that conveys a great deal of power. Works of art are typically granted a certain degree of cultural weight. Describing something as a "work of art" typically implies that one should take certain normative attitudes toward the object. So, for many theorists, it is important that we recognize which things are works of art and understand what it is about works of art that confirms their status. At least since the time of Charles Batteux (1746), the art-status of painting, sculpture, architecture, music, and poetry have been beyond doubt. In the nineteenth century, dance and literature were added to that list. But, how much broader is the domain of art? One can easily find arguments for and against the art-status of numerous different cultural kinds—like debates over the

C. Bartel (✉)
Appalachian State University, Boone, NC, USA

Charles Sturt University, Bathurst, New South Wales, Australia
e-mail: bartelcj@appstate.edu

E. Terrone, V. Tripodi (eds.), *Being and Value in Technology*,
https://doi.org/10.1007/978-3-030-88793-3_7

art-status of food (Korsmeyer, 2002; Telfer, 2008), gardens (Cooper, 2003), perfumery (Brozzo, 2020), screenplays (Nannicelli, 2011), tea ceremonies (Wilson, 2018), and video games (Smuts, 2005; Rough, 2017). It is understandable, then, to wonder whether the things that are referred to as "computer art" count as *art* in this broader, more culturally prestigious sense.[1]

There are a number of routes that theorists may take when seeking to define the boundaries of art. In this chapter, I will explore one route concerning the status of computer art. Some theorists have suggested that the art-status of computer art depends on whether or not it has an artistic medium. I will consider two contrasting views on this topic—those of Timothy Binkley (1988) and Dominic McIver Lopes (2010).

The problem for computer art is to identify what its medium may be and to explain how its possession of a medium may be related to its art-status. While our focus here will be on the second question, theorists have largely focused on the first question. Thus, theorists have wondered whether the medium of computer art is merely a digital version of an older existing artistic medium, or is its medium something distinctive and new. Some technological advancements lead to the development of new media—and therefore of new art forms—while other technological advancements do not. In some cases, digital technologies can offer possibilities for new art forms—video games are not an extension of film, or of anything else. In other cases, digital technologies offer nothing more than new ways to work in old art forms. A digital photograph is still a photograph after all. This is not to say that such technological advancements leave the art world unchanged. New technologies can have a profound effect on an art form without fundamentally changing that art

[1] It may seem strange to question whether computer art counts as *art*. Indeed, one might insist that, because the phrase "computer art" already contains the word "art" within it, then the issue seems to have already been settled. However, we should also notice that the word "art" is sometimes used to refer to things that are not considered to be in the category *art*—consider phrases like "culinary arts", "martial arts", and "the art of conversation". Despite the appearance of the word "art" in these, we may still wonder whether food, fights, and conversation should count as *art* proper. It is also worth noticing that there has been much debate surrounding the art-status of "conceptual art". So, the linguistic use of the word "art" does not offer good guidance for delimiting the concept *art*. I ask the reader to treat "computer art" as a phrase that refers to a certain set of objects, whose art-status is in question.

form or introducing a new medium. This can happen when technological advancements lead to radical changes in the social structures associated with some art form. For example, consider the technology of music recording. For much of the twentieth century, the dominant technology for recording music was analog magnetic tape. Analog tape has a warm tone, but it is both very expensive and quite cumbersome. Today, digital recording technology is far more common in music. This shift from analog to digital has drastically changed the music industry, not by fundamentally changing the practices of sound engineers, but rather by lowering the economic barrier for new people to enter the industry. Many of the old social barriers that have long kept some people out of the music industry have been greatly diminished. With this in mind, we can ask, is computer art a new art form—like video games—or is it a radical new way of participating in an old art form—like music recording?

In what follows, we will first examine two broad approaches to defining "medium", which we can call the physical view and the social-ontological view in Sect. 2. Then in Sect. 3, we will examine Binkley's criticisms of computer art; and we will see how Lopes' account of interactivity is able to address the core criticisms that Binkley offered in Sect. 4. However, in Sect. 5, we will examine two additional problems concerning the notion of media broadly and their impact on the art-status of computer art. The two issues raised in Sect. 5 are inspired by some skeptical worries about the concept of a medium raised by Monroe Beardsley (1981). While it is not my intention to defend Beardsley's skeptical position, I will argue that his skeptical worries raise a problem that is particularly troublesome for theorists who defend the art-status of computer art based on its possession of an artistic medium. Finally, in Sect. 6, I will briefly suggest that a weakened version of Lopes' account may be able to explain our appreciative engagement with computer art, but only if we are willing to abandon our concern for its art-status.

2 Two Approaches to Media

It is reasonable to think that the possession of a medium is a distinctive feature of art. Paying critical attention to the artist's handling of their medium has long been central to art criticism. Often, our artistic evaluation of some work is dependent on our knowledge of the limitations and potentials of the particular medium chosen. Some effects are more easily realized in one medium rather than another, thus appreciating a work may require one to account for the artist's achievement in overcoming whatever difficulties the artist may have faced (Gaut, 2010, p. 293). Additionally, we (typically) pay little regard to the medium of non-art objects (the following argument first appeared in Bartel, 2021). An intuitive suggestion would be that works of art have a medium, while non-art objects do not. This claim is philosophically interesting because often the same materials may be used in either case. Drawings can be rendered in pencil, but so too can diagrams. Consider other examples. Poetry, music, and dance each have a medium while shopping lists, alarm clocks, and jumping jacks do not even though each involve *words*, *sounds*, and *movements* respectively. While it is relevant to the appreciation of art to consider how the artist handles their medium, such concerns appear either irrelevant or frivolous when thinking of non-art objects.

So, what exactly is a "medium"? A *medium*, in its most general sense, is something that stands in-between (Harrison, 2015, p. 294). If we distinguish *what* is done in a work of art from *how* it is done, then claims about the medium are claims concerning the *how*. We can distinguish in the literature narrow and broad definitions of "medium". Some theorists define "medium" narrowly as the physical stuff that works of art are made of. Carroll captures the essence of the narrow definition with his rhetorical question, "[a]ren't media, in the most clear-cut sense, physical…?" (Carroll, 2003, p. 3; see also Beardsley, 1981; Harrison, 2015). This view has some intuitive appeal. Watercolors, pastels, marble, and wood are naturally described as some of the media of painting and sculpture. However, the physical-stuff account of media becomes more complicated when we turn to art forms like poetry, literature, theater, and dance. Alternatively, other philosophers think of the medium as a more complex phenomenon, one that begins with a work's materials, but where the medium is partly constituted by certain socially or culturally sanctioned uses. Richard Wollheim (1987) suggests that we should distinguish between the *material* of art and its *medium* (see also Davies, 2003; Gaut, 2010, p. 288, and Lopes, 2010). Drawings can be made with pencils (among other materials), but pencils can also be used to write poems. So, it cannot be that *pencil* is a medium, one that includes both some drawings and some poems. Instead, *pencil* is a material. According to what we could call the social-ontological view, the medium "is how one uses the material; so the medium is constituted by the set of practices that govern the use of the material" (Gaut, 2010, p. 288). Consider again the case of music recording. The shift from analog to digital has had very little impact on the sanctioned uses of music recordings. Sound engineers who work exclusively with digital recording technologies are doing essentially the same things that older generations of analog engineers were doing. This suggests that the shift from analog to digital does not mark a change of the medium.

Not all philosophers and art theorists are enthusiastic about the notion of the medium however. Beardsley suggested that we might be better off without the concept entirely (Beardsley, 1981, p. 82). Beardsley's skepticism was motivated by two main concerns. First, that the term "medium" is used inconsistently to refer to various qualities that cannot be unified into a single concept; and second, that it is often quite difficult to distinguish whether a claim about the medium is a claim about the work's material or is instead a claim about some observable quality of the work. (This later criticism of Beardsley's will be taken up more fully in Sect. 5 below.) Echoing Beardsley, Noel Carroll has long asserted that the notion of a medium is irrelevant to the appreciation of cinema (Carroll, 2003). More precisely, Carroll argued against claims of "medium specificity", which is sometimes defended as the idea that each art form has a distinctive medium and is sometimes defended as the idea that artists should focus their creative attention on whatever is distinctive of each art form's medium (see Gaut, 2010 for criticism). While these are important debates, I will not comment on them here. Examining debates over medium specificity would result in a significant distraction for our present purposes.

3 Binkley's Criticism of Computer Art

Timothy Binkley's (1988) criticisms of computer art appeared over thirty years ago, and yet the points he raised then still apply today. Binkley does not dismiss the notion that legitimate art can be made on computers, nor does he dismiss "computer art" as a genuine art form. Indeed, Binkley himself is a celebrated computer artist whose work has been exhibited around the world. Instead, Binkley advances the more limited claim that the computer is not a medium of art and can only mimic existing "traditional" artistic media. "Computer art has no characteristic look or behavior, but like a chameleon it can be made to take on the feel and appearance of a multitude of old and new tools" (1988, p. 165).

Binkley offers two arguments that I will focus on here. The first concerns the nature of artistic media. Binkley's account of media seems clearly aligned with the social-ontological camp, but his view is also one that gives prominence to the physical material. Binkley draws on Joseph Margolis' (1984) account, which Binkley describes as having three distinctive characteristics: the medium of a work is "closely connected to a physical material", the work's artistic message is embodied in its physical material, and the medium is a "culturally defined communication channel whose expressive possibilities are determined, at least in part, by the cultural conventions which delimit that channel" (Binkley, 1988, p. 156). While the third point here clearly puts his account of media in the social-ontological camp, Binkley's criticism of computer art focuses on the first point, that of its physical material. Perhaps the best interpretation of Binkley's view is that the possession of a distinctive physical material is a necessary condition for the identification of some medium, which he holds that computer art fails.

Computers are not a medium of art, according to Binkley, because the physical material of computer art is its digital code. As Binkley says, "the numbers which are the 'substance' of computer art are not physical objects but conceptual ones" (Binkley, 1988, p. 157). A work's code will generate displays when the user interacts with it in the right way. The displays are what the user experiences. Whatever artistic value a work has can be attributed to its displays. However, the displays are not the code.

The code always stands at one remove from the user, hidden from direct scrutiny. The "bits" that make up a piece of computer art are never the object of one's attention or the object of artistic value. As Binkley says, "[i]t makes sense to inquire about the size and color of a blob of paint on a palette, but a bit has no characteristic look, sound, or smell because it is a concept, not an object" (Binkley, 1988, p. 158).

There are aspects of Binkley's first criticism that one could quibble with here. *Pace* Binkley, a "bit" seems like the wrong level of analysis. If we wish to construct an analogy between visual computer art and painting, then a bit of digital information would not be analogous to a brush stroke, but rather would be analogous to the chemical composition of the paint. Digital visual artists do not work in bits any more than painters work in chemistry. So, the analogy Binkley seeks to draw between bits of information and blobs of paint is unfair. However, Binkley's first criticism still contains a grain of truth. The important point to take from Binkley here, I suggest, is that the medium of the work is naturally regarded as an aspect of a work that can properly be an object of one's attention. This thought is certainly not implausible. When thinking of traditional artistic media, it would be natural to think that the material aspect of a medium is that which one has a direct experience of, and both the physical and the social-ontological accounts acknowledge the materiality of artistic media. To accept that a piece of software is the medium of computer art, we must reckon with its invisibility.

Binkley's second criticism of computer art concerns its effect on the artist's responsibility: the artist cannot fully be said to be responsible for the look (or sound) of a piece of computer art. Rather, the computer is partly, but significantly, involved in determining a work's final appearance. Computers give artists a powerful tool to manipulate images or sounds in complex ways, manipulations that would be exceedingly time-consuming—and sometimes impossible—to achieve using only analog tools. The computer is not just a tool, but is instead an "intelligent" machine that acts almost like a partner in the creative process. This diminishes the artist's responsibility for the finished product. As Binkley says:

> The medium artist must either create or find each detail. But the computer artist might simply say something like "Put a grove of aspen over here and a spruce forest over there with a meadow in between and bathe the whole setting in a warm afternoon sun." The computer will then determine all the necessary details: where each aspen leaf, spruce needle, and blade of grass is placed and how it reflects the light (Binkley, 1988, p. 162).

Another example from music illustrates this well. An effect that some rock albums employ is the backwards recorded track—an effect called "backmasking". It is difficult to record a backmask track in analog. It requires a great deal of planning and precision. The engineer needs to plot out exactly the start-point and end-point where the backwards track should appear, and then manually flip the analog tape over to record the track backwards. The musician then needs to perform their part, getting the timing right, without recourse to the usual audible cues. Backmasking typically requires a lot of trial-and-error. However, achieving the same effect using a computer is easy. Many recording programs include a function that allows the engineer to select a segment of a track to flip backwards. What would take many hours of concerted effort in analog can be accomplished digitally in minutes.

Backmasking is a localized effect. Other digital effects can be more global. Think of the image filters that have become popular on Instagram. Selecting a filter can radically change the character of an image, often in unpredictable ways. In cases of global effects, the user inputs broad variables into the computer and, as output, gets a detailed piece. The final look of the display will have more nuanced details than the artist could have anticipated. The computer interacts with the user in ways that follow the design of its own programming. It is as if there was an invisible hand acting as a co-contributor to the artist's work. The problem is not that artists fail to be creative, but rather that artists fail to be fully in control. As Binkley describes it, traditional media *react* to the artist's gestures, while computers *interact* (Binkley, 1988, p. 161).

These two points taken together—that computer software is invisible and interactive—poses a challenge for the idea that the computer offers a distinctive artistic medium. Against such challenges, Dominic McIver Lopes (2010) has argued that computer art does have a distinctive medium, which he identifies as its interactivity, and that computer art is a distinctive art form. While Lopes does not identify Binkley directly as one of his interlocutors, his account aims to address the two criticisms discussed above. While there is much that could be said about Lopes' account, the focus of Sect. 4 will be on how his view is able to address Binkley's criticisms.

4 Lopes on Interactivity

Lopes argued that we must first distinguish between *digital art* and *computer art*. Works of computer art form their own distinctive art form while works of digital art, which are created or displayed in a way that is dependent on a computer, are typically continuous with one of the traditional arts—that is, painting, sculpture, music, and so on. For instance, TikTok videos are made specifically for distribution on social media platforms, which makes them dependent for their display on computers. However, such videos do not represent a new art form. Rather, they are a sub-group within the existing art form *moving image*. To classify a work as an instance of computer art, it must do something more. But, what more exactly?

Lopes tells us that computer art is a distinctive "appreciative kind". Lopes has given two different accounts of what it means for something to be an appreciative kind. On his first account, Lopes draws attention to the role that comparisons play in our evaluation and appreciation of art. An "appreciative kind" is a grouping of objects that share some relevant feature such that one could meaningfully compare any member of the group to any other arbitrary member (2010, p. 17). For instance, *punk music* would count as an appreciative kind as one could meaningfully compare different pieces of punk music to each other. Of course, one could meaningfully compare a piece of punk music to any other piece of music—Latin jazz, for instance—but this would only show that both pieces belong to a larger appreciative kind, which is simply *music*. So, the appearance of meaningful comparisons evinces that two works are members of some appreciative kind. In a later work, Lopes gives a slightly different explanation. In his *Beyond Art* (2014), Lopes argues that appreciative kinds are groupings of objects that imply certain appreciative norms. As Lopes says, "[an] art, *K*, is an appreciative kind just in case there is a property of being good qua *K*, or being good-modified for a *K*, or being good qua *K** for a *K*" (2014, p. 146). Again, *punk music* is an appreciative kind because there is such a thing as being good *as a piece of punk music*.

Taking Lopes' two formulations together, we can think of an appreciative kind as a comparison class that is associated with certain appreciative norms. With this in mind, it is worth noticing that, if *digital art* were an appreciative kind, it would need to be a very loose and broad one. There are many different forms of digital art. All that holds them together into a common kind is that they are either made or displayed on a computer. The moniker "digital art" would bind together things like TikTok videos (because their displays are distributed through computers) and mash-ups (because they are made on computers). But it seems absurd to think that many meaningful comparisons could be made between any arbitrarily selected TikTok video and any arbitrarily selected mash-up.

Now, what about *computer art*? Does it signify an appreciative kind? To make this claim, it would be helpful if we could find a distinctive feature that all members of the appreciative kind share, one that offers some grounding for comparisons among members of the kind and for the kind's appreciative norms. To this end, Lopes argues that computer art has a distinctive medium: interactivity. As he says, "computer-based interactivity is the special feature that distinguishes computer art from digital art and indeed all the other arts" (Lopes, 2010, p. 27). So, what is "interactivity"? Unfortunately, various theorists offer conflicting definitions of interactivity, some stronger and some weaker. For instance, David Chalmers defines interactivity quite broadly: "An environment is interactive when actions by the user make a significant difference to what happens in the environment" (Chalmers, 2017, p. 312). By this definition, lots of things are interactive. When I pay my mortgage online through my bank's website, it makes a significant difference to my account. When I put my finger on the glass of a plasma ball, the electric current follows my finger. When I twist the rods of a foosball table, I sometimes get lucky and score a goal. Interactivity abounds.[2]

By contrast, Lopes' account of interactivity identifies its distinctive role in computer art. As Lopes says, "a work of art is interactive just in case it prescribes that the actions of its users help generate its displays" (Lopes, 2010, p. 36). It would be helpful here to pause and consider one of the works that Lopes offers as his main examples. Here is Lopes' description of a piece of computer art by Scott Snibbe, called *Boundary Functions*:

> You and a companion step onto a slight raised platform and a line is projected onto the floor halfway between you. As more people join in, more lines are drawn, creating an irregular tiled pattern. Try as you might—and some children who have just stepped in are trying very hard—you cannot step outside your boundary. Someone remarks that *Boundary Functions* was the title of the doctoral thesis of Theodore Kaczynski, the Unabomber. (Lopes, 2010, p. 25)

Works of computer art generate displays, which are the main object of attention for the users. In *Boundary Functions*, the display is just the irregularly patterned tiles projected on the platform. However, the user themselves must generate the displays through their own input—in this case, it is the act of stepping on the platform that generates the projected lines. This results in a work whose display "changes depending on the actions of its users", its display is "variable" (Lopes, 2010, p. 37). The more users are on the platform, the more lines are projected; and the movement and position of the users will make the resulting tiles change shape. Works of computer art are designed to respond to specific inputs, thus some user actions are prescribed by the work. Additionally, users must be aware that their actions are directly responsible for generating changes to the work's displays—users of *Boundary Functions* come to recognize that they are being divided up into individual tiles. Users explore the work by interacting with it, generating new displays, and seeing what it can do. Through such interactive experimentation, the user comes to understand what

[2] This is not intended as a criticism of Chalmers' account. The definition of interactivity that he offers works perfectly well for his purposes. The point to notice here, however, is that a weak definition of interactivity would not suffice as a distinctive feature for computer art.

other alternative displays are possible. As Lopes says, "[t]hrough many interactions and displays, [users] come to see the possibilities the work holds for them. Seeing these possibilities pulls in something that's not available in each display taken by itself" (Lopes, 2010, p. 60). Users come to understand the work by paying attention to what effect their actions and choices have on the work's displays. In the case of *Boundary Functions*, this means coming to understand how the work artificially separates and contains the individual users. Just as musical works have a formal structure, works of computer art have a structure as well, though the "structure" here is something like an algorithm or a decision-tree that maps out the possibilities afforded by the programming. Users interact with the work as a way of uncovering what cannot be seen—namely, the functional operations of the work's programming.

Taking the above account of computer art, Lopes argues that interactivity itself is the medium of works like *Boundary Functions*. When the user recognizes the work as containing a reactive program and begins piecing together the program's operations, then one is clearly engaging with the underlying structure of the work. Ultimately then, appreciating a work of computer art requires the user to attend to and understand that structure. Perhaps we should still describe the medium as being *invisible*; but nonetheless, when one engages with the work in the prescribed way, the medium of the work is certainly an object of the user's attention. Indeed, considering the kind of interactivity that Lopes describes, the user who engages with the work as prescribed cannot help but to engage with the medium of the work.

At this point, we can see how Lopes' account would address Binkley's two criticisms. Recall, Binkley's first criticism concerned the invisibility of the medium. It is natural to think of the medium of the work as (at least including) the material that the work is made of, but the material of computer art must be its software, and Binkley holds that this cannot be an object of one's attention. Lopes' account offers a compelling response to this criticism: users critically attend to the work's software—in fact, piecing together an understanding of how the software functions is exactly what users are prescribed to do. Software is not visible to the user in a literal sense—users do not make critical comments about particular lines of code—but this does not mean that it is beyond the scope of the user's attention. Just as works of music have a structure that listeners must imaginatively piece together, so too do works of computer art have a structure that users must uncover through their interaction.

Binkley's second criticism was that artists cannot be wholly responsible for their works. The interactivity of computers renders them, not as tools for an artist's craft, but rather as odd silent partners in the creation of a work. However, we can now see that Binkley is thinking about a narrow form of interactivity, one closer to Chalmers' definition. The kind of interactivity that Lopes describes in works like *Boundary Functions* does not limit or diminish the creative input of the artist. While we might describe the computer as partnering with the user to generate the work's displays, there is no reason to think that the work itself—that is, the program—was not fully the responsibility of the artist. Binkley is concerned with cases where a predesigned program alters the artist's input in ways that affect the final look of the work. Instagram filters are a proper target of Binkley's criticism. One creates a piece of visual art *using* a computer, but the user's input is manipulated by the filter's programming to produce the artistic output. However, in the case of *Boundary Functions*, it is the *user's* input that is being manipulated, not the *artist's*.

5 Two Worries

Lopes' account is able to address Binkley's concerns by identifying a form of interactivity that can be a focus of the user's aesthetic and artistic interest. Moreover, Lopes argues that this form of interactivity is the medium of the work, therefore fully refuting Binkely's account. However, I want to suggest that, while Lopes' account takes us a long way toward understanding the art-status of computer art, it leads to yet another problem, one that proves to be intractable.

To begin, we should consider in more detail Beardsley's skepticism about the notion of artistic media. Beardsley claimed that the notion of a work's medium "is now almost useless for serious and careful criticism" (Beardsley, 1981, p. 82). There are two concerns motivating Beardsley's skepticism—the inconsistent use of the term "medium" and the ambiguity between material and aesthetic claims. For our purposes here, I will focus on the second issue. According to Beardsley, it is often quite difficult to distinguish whether a claim about the medium is a claim about the work's material or is instead a claim about some observable quality of the work. The complaint Beardsley makes here is somewhat enigmatic. It would therefore be helpful to quote the paragraph in which it arises in full:

> I take it "This is a water color" is a statement about the medium. Then either the medium is not part of the painting at all, as an aesthetic object, but only its physical basis, or else the medium is simply a set of visible qualities—the thinness, flatness, transparency of the colors. There seems to be no useful way of dividing descriptions of aesthetic objects into those that refer to its medium and those that do not. (82)

Beardsley's view seems to be this: statements about the work's medium may rightly refer to two different sorts of things. Some statements about the work's medium refer to the physical material of the work. However, the physical material of the work is not something that the artist created—it is not "part of the painting". To be "part of the painting" is to be something that the artist created. An object of one's critical, aesthetic engagement should be restricted to things that the artist created. So, statements about the physical material of the work are not relevant to our critical appreciation. Alternatively, other statements about the work's medium refer to observable qualities of the work. Those are certainly things that the artist created. However, "observable qualities of the work" is a broad category that would include things like aesthetic and expressive qualities. Insofar as statements about the medium refer to observable qualities of the work, we cannot distinguish these statements from aesthetic or expressive statements. So, taking all this together, *medium* is a concept that has no distinctive use in art criticism because either it refers to something that is not "part of the painting" or it merely refers to some aesthetic statements.

The view Beardsley espouses here has not been taken up by other philosophers. First, regarding Beardsley's comment that "medium" may refer to the physical material of some work, it is noticeable that he assumed the fairly austere physical conception of media.[3] While some philosophers defend a physical view of media, the dominant view today is some form of a social-ontological view (e.g. Davies, 2003; Gaut, 2010; Lopes, 2010; Margolis, 1984; Wollheim, 1987). Second, few philosophers today would doubt that the critical evaluation of a work should focus on what the artist has made; however, very few philosophers would interpret that quite so narrowly as Beardsley apparently did. The artist's handling of the medium is widely accepted as a legitimate object of critical attention (Davies, 2003, p. 183). Moreover, many would insist that our critical evaluations should additionally take into account factors that fall outside of the narrow scope that Beardsley offers (see Weiser, 2018).

These criticisms of Beardsley aside, there is still an interesting problem that his criticism poses, particularly for computer art. The idea I want to adapt from Beardsley is his claim that, at least in some cases, it is difficult to determine whether a critical statement refers to the medium of the work or instead refers to some quality of the work. Lopes has argued that interactivity is the medium of computer art. His defense of this claim rests on his observation that users are prescribed to experiment with the work to understand how it functions and what the work makes possible. Lopes' account clearly demonstrates that interactivity is a central concern for computer art; but, why should we think that interactivity is the *medium* of computer art? A difficulty arises when we consider other interactive cultural kinds that fail to be art. Works of computer art are interactive, but so too are games. (This is a point that Lopes notes. As Lopes says, "[g]ames are interactive in the sense that they have displays whose features are determined at least in part by their players" (Lopes, 2010, p. 104).) The trouble is that the characteristics of interactivity that Lopes describes in defense of his claim that computer art has a distinctive medium applies equally well to games. Yet, we wouldn't want to conclude that interactivity is the *medium* of soccer. In fact, the thought that *soccer has a medium* seems absurd. Therefore, Lopes cannot rely on his account of interactivity as a piece of evidence in favor of the claim that computer art has a distinctive medium as there are clearly other cultural kinds that

[3] Though some comments in his (1982) seem to suggest that Beardsley later changed his mind. Thanks to an anonymous reader for pointing this out.

are interactive in exactly the same sense but which fail to have a medium. We need some further explanation why interactivity is, in the case of computer art, a medium.

More importantly, I worry that we cannot determine whether an object has an artistic medium until we first decide whether or not that object is a work of art. I suggested earlier that works of art have artistic media while non-art objects do not—at least, this claim seems intuitive. Even if that intuitive claim is false, we may defend the weaker claim that the medium of some object becomes relevant to our critical engagement only when that object is a work of art.[4] If this is correct, then consider again Lopes' account of computer art. While Lopes argues that computer art is a distinctive art form because it has a medium—interactivity—my worry is that we cannot appeal to interactivity as evidence that some object is a piece of computer art because we must first decide whether that object is a work of art. It is only once we recognize something as a work of art that its medium becomes a relevant consideration. But, this is obviously circular.[5]

---

[4] However, even this claim might seem too strong. One might object that the medium becomes a relevant consideration for the appreciation of any object—whether it is art or not—whenever any material is modified or presented with the intention of producing aesthetic properties. (Thanks to an anonymous reader for posing this question.) This suggestion is also intuitively plausible and many cases would seem to confirm it. For instance, one might think that a cook works in the medium of food (this is the reader's example), a fashion designer works in the medium of textiles, and a florist works in the medium of flowers. In each of these cases, some material is manipulated with the intention of producing some aesthetic effect and it is entirely appropriate for the consumer of such objects to turn their attention to the producer's handling of that material. So, there seem to be cases where concern for the medium occurs outside of the domain of art. However, in response to this objection, it is important to notice that such aesthetic activities (cooking, fashion design, flower arranging) could be defended as distinct art forms on the basis that such objects are intentionally produced to achieve certain aesthetic effects. Indeed, many theorists would be tempted to grant art status to objects that are made and presented with the intention of producing aesthetic properties. The upshot of this is that the concept *art* is so bound up with the notion of a medium that we are tempted to call something "art" whenever we turn our aesthetic attention to its material. Thus, I believe we are still justified in considering the weaker claim made here.

[5] I believe the point here may pose a further problem for Lopes' (2014) buck-passing theory of art. If computer art does not have a medium, then it is not *art* on the buck-passing theory; and computer art cannot have a medium unless we already count it as art. So, buck-passers seem unable to account for computer art. (I thank an anonymous reader for this suggestion.) Unfortunately, I do not have the space to consider this line of argument further here, but see my Bartel and Kwong (2021) for critical discussion of Lopes' buck-passing theory.

Boundary Functions is a work of art. It has all the distinctive social cues associated with art. It has a title, its creation is attributed to an artist, it invites interpretation, and it is exhibited in a space where art works are typically exhibited. These social cues suggest that it would be relevant to one's critical engagement to go looking for the work's medium; and once we go looking for the medium of *Boundary Functions*, then interactivity seems like a good candidate. But the important point here is that the work's medium likely becomes an object of critical attention only because we have already decided on independent grounds that *Boundary Functions* is a work of art. We cannot appeal to an object's (apparent) possession of a medium as evidence for its status as a work of art. Instead, the art-status of the work must be explained by appealing to some other factor.

To summarize the two problems raised above, the worries are that (1) interactivity may be a *feature* of computer art even if it is not its medium, and (2) we can only determine whether an object has a medium after we have decided that the object is a work of art. Moreover, there is an alternative hypothesis available to us that can account for the evidence that Lopes provides equally well, which is that interactivity is the primary *mode of engagement* that users employ (and are prescribed to employ) when faced with a work of computer art. Different art forms prescribe different modes of engagement from audience members, the most common being *to look*, *to listen*, or *to read*. Some art forms may prescribe a mode of engagement that requires more than merely looking, listening, or reading. To appreciate sculpture, it may be necessary to look while also moving about the room to see the work from different angles and distances. Other art forms may prescribe more participatory modes of engagement. *Test Site* was an installation work by Carsten Höller in the Tate Modern in London from October 2006 to April 2007.[6] It consisted of a number of twisting slides that museum goers could use. The work's prescribed mode of engagement was apparently *to slide*. Additionally, we should notice that, while there is certainly some relationship between the medium of a work and its mode of engagement, the mode of engagement seems to be the broader category that various artistic media are subsumed

[6] https://www.tate.org.uk/whats-on/tate-modern/exhibition/unilever-series/unilever-series-carsten-holler-test-site

under. Indeed, artistic media are more numerous than are the modes of engagement. Watercolors, oils, acrylic, pencils, inks, pastels, and charcoal are all different media that can be used in picture-making; but the mode of engagement prescribed by each of these media is *to look*.

Consider the way that users engage with *Boundary Functions*. Users step onto a platform and observe the way that the projected lines box in each user in their solitary tile. Users experiment with this when they engage with the work as prescribed. How many people can fit on the platform? How quickly will the lines adapt to the movement of the users? Can the system be fooled and allow one to cohabitate a tile with another user? The mode of engagement for this work requires its users to do a number of things—looking, moving around the platform, and thinking about the underlying behavior of the software. Broadly, we can say that the mode of engagement is *to interact*. This alternative account is able to account for the importance of interactivity in computer art without positing the claim that interactivity is a *medium*. So, the challenge would be to show that Lopes' medium-interpretation of interactivity should be preferred over the mode-interpretation suggested here while avoiding the two worries discussed above.

6 Do We Need *Art*?

Does "computer art" belong to the general category *art*? Some philosophers have sought to address this question by examining whether computer art has a medium. For Binkley, computer art is not a distinctive art form, nor does it have a distinctive medium. Yet computer art is properly part of the category *art*, but only because it imitates existing artistic media. If this is correct, then we should interpret and understand works of computer art simply as technologically produced instances of older art forms—like painting, drawing, sculpture, and music. Against this, Lopes argues that computer art is a distinctive art form, and it is one that has a distinctive medium: interactivity. The interactivity of a work of computer art is a proper object of the user's attention as the understanding and appreciation of works like *Boundary Functions* demands that one come to appreciate its underlying functional structure. I have argued here that, while Lopes' account is able to address the objections raised by Binkley, there is an additional and intractable problem: we cannot identify interactivity as a medium of art until we determine whether the objects in question are works of art. Therefore, the prospect of defending the art-status of "computer art" by appealing to its possession of a distinctive artistic medium is hopelessly circular.

Where should we go from here? While it is certainly possible to pursue some other route to defend the art-status of computer art, in this concluding section I wish to briefly explore the more radical suggestion that computer art might be served perfectly well without the concept *art*.[7] I want to suggest, first, that we should follow Lopes in viewing "computer art" as an appreciative kind. This means that "computer art" is a comparison class where certain normative standards apply. Furthermore, it is an appreciative kind that places high value on interactivity (whether it is regarded as a medium or as a mode of engagement). We can meaningfully compare two arbitrarily selected works of computer art by looking at their use of interactivity and the normative standards we apply in such comparisons take interactivity to be of central importance. "Computer art" is an appreciative kind that is robust enough to sustain the user's aesthetic and evaluative engagement with individual members of the class. Indeed, the notion that computer art is an appreciative kind is a useful one.

However, I suggest that nothing of aesthetic or evaluative significance follows from its status (or not) as a work of art. What would change about our engagement with something like *Boundary Functions* if we decided that it was not a work of art? Would we refuse to step onto its platform? Would we pay little regard to the lines it projects? Would we no longer compare it to other interactive computer programs? I doubt it. In fact, I suspect that nothing about our engagement and evaluation of *Boundary Functions* would change at all.

This is not to say that the way in which we categorize the piece has nothing to do with its aesthetic value. In a highly influential essay, Kendall Walton (1970) argued that categories play both a psychological and normative role in criticism. One of Walton's key claims is that the way in

[7] For a brief defense of art eliminativism, see Bartel and Kwong (2021).

which we categorize an object influences how we aesthetically respond to it. However, I suggest that the kinds of categories that matter are categories like appreciative kinds, and not categories like *art*. Everything that we find aesthetically rich and meaningful about our engagement with *Boundary Functions* comes from its status as an appreciative kind. Indeed, it is likely that aesthetic and evaluative norms are generated by categories at the level of appreciative kinds. By contrast, *art* itself is not an appreciative kind. To see this point, remember that appreciative kinds are classes associated with certain appreciative norms where one should be able to draw meaningful comparisons to any two arbitrarily selected members of the class, and *art* clearly does not satisfy this definition. *Art* is too broad of a category to hope that meaningful comparisons could be drawn between any two arbitrarily selected members, and the appreciative norms of *art* are too varied and inconsistent to identify with a single class.

The purported aim of the debate that was examined here is to establish ultimately whether computer art is *art*; however, I have argued that we cannot settle that debate by appealing to its possession of an artistic medium. Despite this, the debate over the medium of computer art is not a wasted debate, but rather it serves a different function. What the debate helpfully does is to establish some of the normative standards of computer art as an appreciative kind and to identify the central role of interactivity. But in appreciating "computer art", we may not need a concept of *art* at all.

References

Bartel, C. (2021). Art, Aesthetics, and the Medium: Comments for Nguyen on the Art-Status of Games. *Journal of the Philosophy of Sport*, https://doi.org/10.1080/00948705.2021.1948336

Bartel, C., & Kwong, J. M. C. (2021). Pluralism, Eliminativism, and the Definition of Art. *Estetika: The European Journal of Aesthetics, 58*, 100–113.

Batteux, C. (1746). *Les Beaux-arts Reduits à un Même Principe*. Durand.

Beardsley, M. (1981). *Aesthetics: Problems in the Philosophy of Criticism* (2nd ed.). Hackett.

Beardsley, M. (1982). What is Going on in a Dance? *Dance Research Journal, 15*, 31–36.

Binkley, T. (1988). The Computer is Not a Medium. *Philosophic Exchange, 19*, 155–173.

Brozzo, C. (2020). Are Some Perfumes Works of Art? *Journal of Aesthetics and Art Criticism, 78*, 21–32.

Carroll, N. (2003). *Engaging the Moving Image*. Yale University Press.

Chalmers, D. (2017). The Virtual and the Real. *Disputatio, 9*, 309–352.

Cooper, D. (2003). In Praise of Gardens. *British Journal of Aesthetics, 43*, 101–113.

Davies, D. (2003). Medium in Art. In J. Levinson (Ed.), *The Oxford Handbook of Aesthetics* (pp. 181–191). Oxford University Press.

Gaut, B. (2010). *A Philosophy of Cinematic Art*. Cambridge University Press.

Harrison, A. (2015). Medium. In M. Kelly (Ed.), *Encyclopedia of Aesthetics* (2nd ed.). Oxford University Press.

Korsmeyer, C. (2002). *Making Sense of Taste*. Cornell University Press.

Lopes, D. M. I. (2010). *A Philosophy of Computer Art*. Routledge.

Lopes, D. M. I. (2014). *Beyond Art*. Oxford University Press.

Margolis, J. (1984). Film as Fine Art. *Millennium Film Journal, 14*(15), 89–104.

Nannicelli, T. (2011). Why Can't Screenplays be Works of Art? *Journal of Aesthetics and Art Criticism, 69*, 405–414.

Rough, B. (2017). The Incompatibility of Games and Artworks. *Journal of the Philosophy of Games, 1*. https://journals.uio.no/JPG/article/view/2736

Smuts, A. (2005). Are Video Games Art? *Contemporary Aesthetics, 3*. https://digitalcommons.risd.edu/liberalarts_contempaesthetics/vol3/iss1/6/

Telfer, E. (2008). Food as Art. In A. Neill & A. Ridley (Eds.), *Arguing About Art: Contemporary Philosophical Debates, ed* (pp. 11–29). Routledge.

Walton, K. (1970). Categories of Art. *Philosophical Review, 79*, 334–367.

Weiser, P. B. (2018). Feminism in Context. In D. Goldblatt, L. B. Brown, & S. Patridge (Eds.), *Aesthetics: A Reader in Philosophy of the Arts* (pp. 524–531). Routledge.

Wilson, D. P. (2018). The Japanese Tea Ceremony and Pancultural Definitions of Art. *Journal of Aesthetics and Art Criticism, 76*, 33–44.

Wollheim, R. (1987). *Painting as an Art*. Princeton University Press.

Breaking the Fourth Wall in Videogames

Nele Van de Mosselaer

"Part of the early work in any medium", Janet Murray writes, "is the exploration of the border between the representational world and the actual world" (1997, p. 103). Referring specifically to people's inability or even aversion to cope with increasingly elaborate simulations of (virtual) reality, she adds that we always need "time to get used to any increase in representational power. During this time one of our main activities, as creators and audience, involves testing for the boundaries of the liminal world" (Murray, 1997, p. 103).

Although Murray specifically talks about difficulties in distinguishing the actual from the represented in our interactions with any relatively new, representational medium, the issues she raises are still relevant with regard to videogames, which can hardly be described as "new" anymore. Philosophers and game scholars are still unable to reach a consensus on the "fictionality" or "actuality" of (our interactions with) the virtual environments, characters, and events that can be found within videogames

N. Van de Mosselaer (✉)
University of Antwerp, Antwerp, Belgium
e-mail: Nele.VandeMosselaer@uantwerpen.be

E. Terrone, V. Tripodi (eds.), *Being and Value in Technology*,
https://doi.org/10.1007/978-3-030-88793-3_8

(cf. Aarseth, 2007; Sageng, 2012; Robson & Meskin, 2016; Chalmers, 2017; McDonnell & Wildman, 2019). At the heart of the issue is the specifically interactive way in which digital technologies allow us to represent and experience fictional worlds. The way in which videogames allow for interactions between actual players and fictional characters at first sight seems to defy the inherent ontological barrier that delineates the actual world from merely represented, fictional worlds (cf. Van de Mosselaer, 2018a).

This chapter discusses how digital technologies complicate the distinction between the actual world and fictional gameworlds by focusing on situations in which what Murray calls "the boundaries of the liminal world" are foregrounded (1997, p. 103). More specifically, I will discuss so-called "fourth wall breaks" within the experience of fictional gameworlds. For this purpose, I will first define the videogame experience as a self-involving, interactive fiction experience, based on Kendall Walton's account of fiction (1990). Afterwards, I will describe how the inherent interactivity and self-referentiality of videogame fiction might seem to complicate the conceptualization of the fourth wall within the videogame context, but actually offers new ways of breaking this imaginary wall.

1 Virtually Representing Fictional Worlds

Before I discuss the boundaries between the actual world and fictional gameworlds, I want to describe how videogames represent such fictional worlds. For this, I will make use of Kendall Walton's make-believe theory of fiction. I will first give a general sketch of this theory and then apply it specifically to videogame fictions.

In *Mimesis as Make-Believe*, Walton writes that something is a fiction when it has the function of serving as a prop in a game of make-believe (1990, p. 51). Make-believe games are, according to Walton, the key to understanding fiction. He models his entire fiction theory on the make-believe games children play. A famous example he describes is that of two boys, Eric and Gregory, playing a game in which they pretend that every tree stump they see is actually a bear (1990, p. 37). Eric and Gregory agreed on a certain principle of generation ("all tree stumps are bears"), according to which fictional truths are generated based on real features of actual objects. The tree stumps function as props: objects that truly exist, but prompt certain imaginings within the game of make-believe. Such props make sure that the imaginings of people interacting with these objects are not just free-floating fantasies: they are both structured and constrained by the features of the prop (1990, pp. 38–39). When Eric and Gregory observe a particularly big tree stump, for example, they will imagine as a result that there is a big bear in their vicinity. Moreover, the make-believe game played by Eric and Gregory is a reflexive one: they themselves also become props in this game. Whenever they run away from a tree stump, they imagine of themselves that they are running away from a bear.

Walton says that the appreciation of representations or fictions (he uses these terms interchangeably) involves precisely these kinds of constrained make-believe games. To understand how we appreciate representational works of art, Walton says we need to look at the specific way in which we imaginatively participate in these works (1990, p. 208). When reading a novel, watching a play, or looking at a painting, for example, we play make-believe games in which we use the novel, the actions of the actors, and the painting as props. The entirety of truths that are generated within such a make-believe game, form what Walton calls a "fictional world" (1978a, 10). In this process, certain principles of generation are always in play and structure our imaginings. Just like the size of the tree stumps determined the size of the imagined bears, the objective features of representations determine what kind of imaginings are appropriate when appreciating them. Moreover, just like in Eric and Gregory's game, the make-believe games played by appreciators of works of fiction like novels and movies are reflexive or self-involving according to Walton:

> What is not so obvious, but of very considerable importance, is that viewers and readers are reflexive props in these games, that they generate fictional truths about themselves. Many of their actions, like those of participants in children's games, are reflexive props as well. And as in the case of participants in children's games, it is in a first-person manner that appreciators are to, and do, imagine about themselves; they imagine, from the inside, doing things and undergoing experiences. (Walton, 1990, pp. 213–214)

One famous example Walton gives us of such appreciator involvement is that of Charles, who watches a horror movie about a terrible, green slime monster. Walton describes how the green slime oozes across the screen, towards the camera. As a consequence of this, Charles imagines of himself that he sees the slime and that he is in danger of being attacked by this slime (Walton, 1990, p. 242). Charles himself thus becomes a reflexive prop in his game of make-believe. He generates fictional truths about himself (so-called "de se imaginings"), such as being part of the world depicted in the movie, being threatened by the slime, and being fictionally afraid (Walton, 1990, p. 242). Walton claims that all fiction appreciators are similarly involved in the fictional worlds they experience through work of fiction. Appreciating fiction, he writes, always entails participating in it (1990, p. 213).

Walton's idea that appreciators of fiction are always also participants in the fiction has been heavily criticized within the philosophy of fiction. Derek Matravers, for example, denies that representations necessarily involve their audience in the world they depict. He points out that, while Walton uses the examples of the reflexive make-believe games played by Charles, Gregory, and Eric as an archetype to explain all of our interactions with fictions, these examples are actually more of an exception (Matravers, 2014, p. 116). After all, in most of our interactions with fictional works, such as paintings, movies, and novels, we are not involved in the represented content whatsoever. Only when a work asks us to involve ourselves in the world it depicts, it can be said to mandate reflexive make-believe games. This is the case when, for example, there are asides (dramatic situations in which characters communicate directly to the audience) or other instances of the fourth wall being broken, through which the audience becomes involved in the fiction. As Walton believes that we are always part of the imagined world when experiencing fiction, his theory seems to imply that asides to the audience or a breaking of the fourth wall are "intrinsic to engaging with representations" (Matravers, 2014, p. 116). Most works of fiction, however, do not force us to be part of the fictional world they represent. The *Green Slime* is an exception because the movie gives the impression that the slime is looking and coming at the spectator, so that it becomes fictional that spectators are in danger. However, unless there is such a breaking of the fourth wall, there is no prompt at all for appreciators of movies, novels, or plays to imagine of themselves that they are watching or somehow involved in the depicted content. They are simply not part of the make-believe game that needs to be played to interpret this content (cf. Currie, 1995, p. 179). Walton's theory seems unable to make the distinction between a detached and a participatory way of appreciating fictions, as his make-believe theory already assumes that "[a]ppreciation of representational works of art is primarily a matter of participation" (Walton, 1990, p. 213).

This aspect of Walton's make-believe theory proved to be generally unpopular among scholars investigating our experiences of non-interactive representational works such as movies and novels. Yet, his emphasis on participation in appreciating representations is also the reason why Walton's theory was extensively applied to videogames (cf. Tavinor, 2009; Bateman, 2011; Robson & Meskin, 2012; McDonnell & Wildman, 2019). "Given that participating with videogames is also primarily an act of engaging with a representational prop," Tavinor writes, "[Walton's] theory of fiction is entirely apt to capturing the nature of the fictive practice involved in videogames" (Tavinor, 2005, p. 30). Chris Bateman remarks that Walton's descriptions of the make-believe games played by readers of novels and viewers of movies gesture at something "players of digital games are intimately familiar with", since videogames allow us to enter their fictional world and experience it from the inside (Bateman, 2011, p. 167). Within the videogame experience, digital technologies are employed as props mandating reflexive make-believe games (cf. McDonnell & Wildman, 2019). The videogame player is cued into imagining not only a fictional world based on what appears on the screen, but also their own involvement and interactions within this world. As Robson and Meskin write, videogame players are, just like Charles in the make-believe game that Walton describes, "almost invariably characters in the fictional worlds associated with video games" (Robson & Meskin, 2016, p. 167).

Videogames offer specific props for players to imagine this kind of self-involvement. Through the figure of the avatar, for example, players are invited to imaginatively identify with a character in the fictional world (cf. Tavinor, 2009, p. 70; Robson & Meskin, 2016, p. 168). Haptic feedback of input devices can help in this identification process: when the avatar is hit by bullets, for example, players can be made to feel this through the rumbling of the controller. Moreover, players also use the real actions they perform on these controllers, and the graphical representations they thereby create on the screen, as props to imagine that they are actually performing actions within the fictional world (cf. Van de Mosselaer, 2018b). Videogame manuals and interfaces are even made to reflexively explain how the player's actual actions are related to the fictional actions of their avatar in the game world. They spell out what Walton calls the principles of generation of fictional truths (1990, p. 38). They might clarify, for example, the relation between "pressing X" and "jumping", which shows how players can generate the fictional truth that they are jumping in the fictional world by performing an action in the actual world. In other words, the videogame experience is a prime example of how digital technologies can be used as cues for de se imagination.

2 A Virtual, Fourth Wall?

Having discussed how videogames represent interactive, fictional worlds, I now want to focus on the imaginary boundary of these worlds: the so-called fourth wall. The concept of the fourth wall was originally used with regard to theatre (Diderot, 1947). It referred to the convention of imagining an invisible wall between the audience and the fictional world represented on the stage. Through this wall, the audience peeks into the fictional world of the play, while remaining unseen by the characters. A fourth wall break happens when the characters can suddenly see through this wall too and somehow acknowledge the world beyond it or the artificiality of their own world.

Nowadays, the concept of fourth wall breaks is used to refer to any instance in which fictional characters somehow prove to have knowledge about the world outside of their fiction (cf. Conway, 2010, p. 147). There are two especially common ways in which works of fiction break the fourth wall. First of all, characters can break the fourth wall by addressing or somehow interacting with the appreciator of the fictional work they are part of. In such instances, this appreciator is no longer a passive observer, but is involved in the fictional world and the characters' adventures in it. As Walton puts it, such fourth wall breaks make for make-believe moments of interaction between the appreciator and one or more of the work's characters (Walton, 1990, p. 235). Secondly, works of fiction can break the fourth wall by self-referentially revealing their own artificial, fictional, or mediated nature. They can do so by having characters refer to their author, address the fact that they are part of a fictional narrative, or acknowledge the technological apparatus supporting the fictional world (such as the camera or boom microphones in movies) (Conway, 2010, p. 147). By doing this, such works acknowledge, from within their narrative, their own status as fictional narratives made-up for actual audiences to enjoy.

Many game scholars have discerned a problem when it comes to applying the concept of fourth wall breaks to videogames. Videogames are interactive, representational works that are inherently characterized by involvement of their appreciators, which is usually taken as a characteristic of works that break the fourth wall. The concept of fourth wall breaks originally used to refer to those situations in theatre where characters are suddenly able to observe and engage with the audience. Conway points out how this description of fourth wall breaks poses a problem for their occurrence in videogames, as videogame players simultaneously "fulfil the dual role of audience member and performer on stage" (2010, p. 146). The distance that previously existed between the audience in the actual world and the characters in the fictional world is, within the videogame experience, thus diminished to the point where one person, the player, exists in the two worlds at the same time. Videogames inherently allow players to take on a fictional role and influence the fictional events in the worlds they represent. Moreover, this kind of interaction presupposes a kind of self-referentiality. Videogames make it clear to their players how they can and should interact with their software and hardware and they typically do this in a metareferential way that reveals their own mediality. "As interactive systems", Jørgensen writes, "digital games must provide system information to the player, and one way to do so in a clear and unambiguous manner is to give extra emphasis to the mediation process itself" (2013, p. 125). Jørgensen here not only refers to the many symbols that appear on the screen and overlay the fictional world, such as health bars and indications of which button the player should press next, but also to game characters who, for example, state that they are "overburdened" when a player picks up too many items. Such metareferences and reflexive statements, Jørgensen concludes, are inherent to "the way games work" (2013, p. 126).

This inherent interactivity and self-referentiality of videogames has led many game scholars to claim that the concept of the "fourth wall" is not applicable to our experiences of fictional gameworlds. The existence of a fourth wall in videogames, or at least one that is intact, is therefore often altogether rejected. Klevjer, for example, claims that "game fictions are not delineated by a 'fourth wall' as in film or literature" (2006, p. 59). Jørgensen agrees, stating that "[t]he presence of a player who has the agency to affect the content of games necessitates a constant breaking of any imagined wall that separates the game reality from the actual world" (2013, p. 125). Similarly, Brown writes that "the fourth wall's position is already compromised by the control system or interaction device's existing outside of the diegesis" (2012, p. 163). Weise, in this regard, describes the technologies that make the representation of gameworlds possible as "umbilical cords", connecting actual players with fictional worlds (2008). According to him, it is more useful "to think about the boundary between player and fiction as an elastic membrane" rather than a wall (Weise, 2008). Kubiński concludes that it would be reasonable to say that "the concept of 'breaking the fourth wall' is not quite accurate in the context of video games" (2014, p. 135).

3 Misconceptions About the Fourth Wall

Before discussing my take on fourth wall breaks in videogame fiction, I want to address some of the ambiguities and misconceptions that can at times confuse discussions about this subject. The first one concerns the kind of "interaction" between actual people and fictional worlds that is enabled by fourth wall breaks. Some work on fourth wall breaks seems to suggest that these breaks allow actual connections to form between the actual and the fictional world, and that they actually make it possible for real people to interact with fictional entities. Klevjer, for example, explicitly talks about the "fourth wall" as "the ontological boundary of fiction"

(2006, pp. 113–114). If understood like this, however, it would become necessarily and inherently impossible to "break" this barrier. Kendall Walton describes how cross-world interaction between the actual world and fictional world is, by definition, impossible, because of the "logical or metaphysical barrier between them" (1978b, p. 11). Peter Lamarque stresses that there is a logical gap between the actual world and fictional worlds, and "fictional characters as such can never cross these logical barriers" (Lamarque, 1981, p. 299). Alex Neill talks about an ontological gap between fiction and reality, which precludes any rivalry or physical interaction between us and fictional characters (1993, p. 4). Indeed, how would a reader of *Harry Potter* even try to warn the wizarding world of an impending Death Eater attack? How would these Death Eaters ever be able to hurt this reader? As we belong to vastly different planes of existence, such endeavours are simply impossible.

Note that this is even true for videogames: we do not *actually* fight aliens when playing *Mass Effect* 2 (BioWare, 2011), but are only mandated to imagine fighting aliens (cf. Van de Mosselaer, 2018a). Likewise, I have never truly shot a zombie, as these monsters simply do not exist. In the guise of Ellie, the player-character in *The Last of Us Part II* (Naughty Dog, 2020), however, I have fictionally shot hundreds of zombies. And with that I mean that I have pressed the R2 button hundreds of times, causing representations of dying, infected creatures on my screen. In other words, the best fictional works can do is mandate appreciators *to imagine* that the barrier between the actual world and the fictional world is overcome. Digital technologies are specifically apt to do this, as they can offer the player very advanced props (input devices, haptic feedback, VR headsets that show a perspective on a gameworld based on the player's head movements, etc.) for imagining their presence within a represented world. But the ontological barrier between worlds remains unbroken, and what Walton writes remains true even for videogames: "What happens in the fictional world—what fictionally is the case—can indeed be affected by what happens in the real world. But one person can save another only if they live in the same world. Cross-world saving is ruled out, and for similar reasons so is cross-world killing, congratulating, hand-shaking, and so forth" (Walton, 1990, p. 195).

Understanding fourth wall breaks as breaks of the "ontological boundary of fiction" is thus quite problematic, as it seems to suggest that fourth wall breaks actually allow fictional characters to cross the boundary to the appreciator's world, or vice versa. In *Mimesis as Make-Believe,* Walton specifies that fourth wall breaks only "bring the appreciator into the fiction" in the sense that it becomes *fictional* "that the reader or spectator is noticed or addressed or offered a drink or threatened" (Walton, 1990, p. 233). That is: fourth wall breaks make it fictionally true of the appreciator that they live in the same world as the fictional characters, thus making some fictional interactions possible between those characters and the (fictional version of) the appreciator.

In light of this, Conway's "reformulation" of the fourth wall in videogame contexts seems hardly original. He writes that games have the unique potential to "not only break the fourth wall, but to expand it, relocating it entirely behind the player, as a tool of immersion" (2010, p. 153). He describes examples of "how the videogame complex does not break the fourth wall, but instead expands the fourth wall, as the player is now placed by the designer within the boundaries of the game fiction" (2010, p. 147). Yet, as discussed in the previous paragraph, this is simply what always happens when a fourth wall break occurs: the fictional world is expanded so as to include the viewer or spectator, of whom it becomes fictionally true that they are somehow interacting with a character.

Secondly, the frequent claims that fourth wall breaks are hard or even impossible to conceptualize within videogame contexts also seem to be misguided. As mentioned before, scholars often argue that, if it is true that videogames always represent fictional worlds in a way that is self-referential and involves the appreciator, then the fourth wall is either never present (Klevjer, 2006, p. 56) or always broken (Jørgensen, 2013, p. 125). Although Walton did not treat videogames in his analysis of representational works, he already addressed the problem of what "fourth wall breaks" (in the form of "asides") might mean for fictional works that already involve their appreciators within the world they represent.[1] Thus,

[1] As discussed before, Walton was convinced that all experiences of fiction involve de se imaginings: an appreciator of a fictional work is always also a participant in the fictional world represented by this work. Even if we do not agree with this thesis, what Walton says is certainly true for interactive fictional media such as videogames. Also, even though Walton indeed did not analyse videogames

Walton asks: if we already belong to the world of a fiction anyway, "why should there be anything special or remarkable about asides?" (1990, p. 233). The answer Walton gives to this question applies equally well to fourth wall breaks in videogames:

> Part of the answer lies simply in the fact that being recognized or addressed in real life marks a significant change in one's social situation. (Think of a student suddenly called on in class or a lecturer picking out someone in the upper gallery and speaking specifically to him.) One feels included in a manner one wasn't previously. An aside makes it fictional that the appreciator is included similarly. This change may be important, but it doesn't consist in the appreciator's suddenly being drawn into a fictional world to which until then he did not belong. (Walton, 1990, p. 233)

Thus, fourth wall breaks do not occur when works of fiction simply involve the appreciator in the represented world, but rather when they do so in a manner that is new or unexpected, because it breaks with the previously established conventional relation between the real and the fictional.

Having discussed these misconceptions about the fourth wall, we can conclude that the fourth wall has often been misconstrued as an occurrence in which the ontological boundary between the fictional and the actual world is somehow broken. Neither fourth wall breaks nor interactive works of fiction can break such an ontological boundary, however, but can at most mandate the imagining of cross-world interactions. Fourth wall breaks are thus not ontological breaks, but rather phenomenological ones: they change the way in which an individual appreciator experiences the world represented in the work. They are moments in which the work of fiction refers to its own mediality or involves their appreciators in the fictional world in a way that is new or unexpected by the appreciator based on the conventions of the particular medium in which they occur. This means that the interactivity and self-referentiality that is inherent to interactive fictional media such as videogames does not automatically entail a constant breaking or a total absence of the fourth

---

in *Mimesis as Make-Believe*, he does mention "interactive fiction for computers" once in this work, in a footnote to his discussion of fourth wall breaks (1990, p. 235).

wall. And this, in turn, means that fourth wall breaks need not be rephrased or even deemed impossible to apply to videogames. The next part of this chapter will thus be a discussion of videogame situations that can be deemed fourth wall breaks.

4 Fourth Wall Breaks in Videogames

As Murray writes, investigations of fourth wall breaks are explorations of the boundary conventions of the representational medium in which the fourth wall break occurs (1997, p. 103). If we want to discuss fourth wall breaks with regard to videogames, we must look at situations in which the conventional interaction with the fictional gameworld is somehow disrupted. As discussed before, this conventional interaction always somehow entails metareferences or in-game representations that reveal the game's mediality and how it is supposed to work. Such self-references are not unexpected curiosities, but rather conventional aspects of how games work (Jørgensen, 2013, p. 126). Conventional interactions with videogames also entail self-involvement of the player. Players take control over an avatar or proxy within the gameworld and can influence fictional events through this figure. Whatever fourth wall videogame players imagine there to be, it is not one that would impede them from being fictionally involved in the gameworld in this way.

Fourth wall breaks thus occur in videogames when the conventional interaction with fictional gameworlds, and the self-referentiality and self-involvement this entails, are somehow disrupted. Examples of this abound. First of all, even though the agency of the player, in the guise of the player-character, does not entail any break of the fourth wall, creative ways of representing the relation between player and avatar can cause such a break. Conway, in this regard, describes how the character Sonic in the game *Sonic the Hedgehog* (Sega AM3, 1993) will get impatient with the player if they do not press any buttons for a long time. Sonic will then "cross his arms and tap his feet, gesturing at the player in frustration, before eventually walking off-screen, resulting in a game-over (Conway, 2010, p. 148). What makes this a fourth wall break is not the fact that the player controls a character within the fiction, as this is conventional, but the fact that this character acknowledges the fact that he is being controlled by the player.

Secondly, fourth wall breaks can happen through unusual uses of interface elements. As already discussed, self-referential interface elements that give the player necessary information on how the game works do not necessarily entail fourth wall breaks. Even though their presence on the screen is a constant reminder of the game's mediality and its status as a mere game, they are not usually experienced as disruptive by players (cf. Jørgensen, 2012). Rather, they are quite unproblematically interpreted as giving information about the fictional world, but not belonging to this world themselves. Health bars, for example, mandate players to imagine their character being healthy or rather close to dying, but do not mandate players to imagine that an actual bar is floating above the heads of their character within the fictional world. Thus, a fourth wall break can happen when this conventional use of interface elements is somehow disrupted. This happens, for example, in *Kingdom Hearts II* (Square Enix, 2005), when the character Stitch quite literally jumps on the player-character Sora's health bar and starts licking it, "healing" Sora in the process.

Thirdly, even though players are used to interacting with game systems when exploring fictional gameworlds, the technological apparatus supporting the videogame experience can be used for breaking the fourth wall. This happens, for example, when Mr. Resetti in *Animal Crossing* (Nintendo EAD, 2001) threatens to delete the player's save file, as if the fictional character is aware of the fact that his entire world is contained into a file on a computer. Game designers can also deliberately make it seems as if their game malfunctions or crashes by having the game show visual anomalies. Such digital artefacts that resemble the visual aspects of real computer glitches are also called "glitchalikes" (cf. Moradi, 2004, p. 10; Gualeni, 2019). One of such glitchalikes appears in *Batman: Arkham Asylum* (Rocksteady Studios, 2009) when Batman meets his enemy Scarecrow. Upon approaching Scarecrow, the screen suddenly freezes and the game seems to glitch. By using such a glitchalike, developers foreground and disrupt the media-fiction relationship that is inherent to videogames, but that is often taken for granted by players as a conventional part of gameplay.

Another exceptionally creative example of using the computer-mediation of videogames to break the fourth wall can be found in the game *Undertale* (Fox, 2015). At the beginning of this game, players are forced to fight the lovely and motherly character Toriel to be able to proceed. After killing her, they meet the game's antagonist Flowey, who informs them that they are a monster for choosing to kill Toriel, while they could have spared her. Quite often, players react to this information by quitting the game, going back to a previous checkpoint on their save-file, and repeating the encounter with Toriel, this time sparing her. When they meet Flowey after letting Toriel live, however, he says the following;

> Clever, verrrryyy clever. You think you're really smart, don't you? In this world, it's kill or be killed. So you were able to play by your own rules. You spared the life of a single person. Hee hee hee…But don't act so cocky. I know what you did. You murdered her. And then you went back, because you regretted it. Ha ha ha ha… You naïve idiot. (Fox, 2015)

While it is unproblematic that Flowey addresses the fact that the player killed Toriel (as this was an in-game event), it is very unconventional that he can also comment on their actions outside of the game. Flowey apparently knows about the player quitting the game and opening a previous state of their savefile, which are things that happened outside of the world of *Undertale*. Flowey thus breaks the fourth wall.

Although many more examples could be discussed here, we can already conclude that fourth wall breaks in videogames do not happen constantly, even though videogames are inherently self-referential and interactive. They only happen when this self-referentiality and interactivity occur in a way that breaks with previously established conventions of the medium.

5 A Unique Fourth Wall?

Many scholars have called the specific ways in which videogames can break the fourth wall "unique" or "complex" because of the way the boundaries between the fictional and the real are always blurred within videogame play (cf. Klevjer, 2006; Weise, 2008; Conway, 2010; Jørgensen,

2013; Kubiński, 2014). Yet, many of their arguments seem to be supported by a misunderstanding about the concept of the fourth wall. As I argued, fourth wall breaks in videogames, just like in all other (non-interactive) representational artefacts, consist of a break in the *conventional* relation between the real and the fictional world. Even though this conventional relation between the real and the fictional might be different when interacting with virtually represented worlds than when experiencing the fictional worlds described in novels or shown in movies, the fourth wall can still be defined based on this relation. As Murray says, every medium simply has its own "boundary conventions" that will decide when a fourth wall break occurs (1997, p. 103).

That does not mean that there is not anything unique about fourth wall breaks that can occur within videogames, however. Up until now, both within this chapter and wider academic discourse, fourth wall breaks have mainly been discussed as disruptive events that happen to an audience when they appreciate a work of fiction. Works of fiction that feature fourth wall breaks represent fictional events in such a way that their appreciators will be invited to imagine, at a point, that they and the characters described in the work are part of the same world (either because the character addresses them or mentions their own artificial nature). In this regard, fourth wall breaks have been understood as disruptive elements that are inherently present in fictional works and that audiences simply come across while appreciating these works.

In the last part of this chapter, I want to show how appreciators can take on a more active role in the occurrence of fourth wall breaks in videogames. Moreover, I want to suggest the possibility of fourth wall breaks that are initiated not from within the work or by the work's creator, but rather by the mischievous player.

6 Two-Directional Fourth Wall Breaks

Whereas non-interactive media very often have fictional characters address actual appreciators, causing the fourth wall to crumble, the reverse seems impossible: when it comes to appreciators trying to address (or interact in any way with) fictional characters in novels or movies, the wall seems indestructible. As interactive fictional media, however, videogames allow for breaks of the fourth wall that are more interactive in nature. In this regard, Agata Waszkiewicz makes a distinction between one-directional and two-directional fourth wall breaks within videogames. While one-directional fourth wall breaks consist in fictional characters who address the appreciator (and are thus the kind of breaks that can be found even in non-interactive fictional media), two-directional breaks are situations in which the player is not only addressed, but also invited to participate (2020). Waszkiewicz describes the latter using an example from the game *Metal Gear Solid* (Konami Computer Entertainment Japan, 1998). When meeting the enemy Psycho Mantis in this game, this character wants to prove his psychokinetic power and asks the player to put their controller on the floor. He then waves his hand across the screen, "moving" the controller across the floor by making use of its rumble function. Afterwards, Psycho Mantis faces the player in a battle that can ultimately be won by the player if they unplug their controller and put it in the "Player Two" slot. Psycho Mantis then cries out that he is now unable to read the player's mind. The fourth wall break is thus two-directional: while the player is addressed by Psycho Mantis, they also have the power and means to answer, not just within the gameworld and in the guise of a player-character, but as themselves, through interaction with their actual console (Waszkiewicz, 2020).

Walton already mentioned this possibility of such "two directional" fourth wall breaks when talking about the interactive moments created by asides. As he focuses primarily on non-interactive fictional media such as literature and film, however, he emphasizes the severe limitations that are usually inherent to such interactions:

> It is rarely fictional that the appreciator carries on an extended conversation with a character, or that he and a character make eyes at each other or exchange more than a glance or two. There is an obvious practical reason for these limits. When the artist constructs her work, she determines, once and for all, what fictionally her characters say and do. But different appreciators will behave differently in front of the work; what fictionally they say and do, what they choose to attend to and how, what they mutter under their breath will vary greatly, and some will behave in ways the artist did

> not foresee. So the artist cannot fit her characters' responses to what, fictionally, the appreciator says or does. (Walton, 1990, p. 235)

Similar to Waszkiewicz, Walton then acknowledges that at least two kinds of fictional media can perfectly be customized to account for multiple possible reactions of their appreciators: improvisational theatre and interactive fiction for computers (Walton, 1990, p. 235). Although Walton does not elaborate on this, it is easy to see how actors in improvisational theatre can respond to whatever reactions they receive from the audience during an aside. Videogames, moreover, can achieve the same effect through the careful programming of their software. Fourth wall breaks, as a result, can become prolonged moments of interaction, instead of mere abrupt disruptions of fiction that happen to appreciators.

7 Appreciator-Initiated Fourth Wall Breaks

Just like all the previous examples of one-directional fourth wall breaks, the two-directional fourth wall breaks that were just discussed are explicitly designed into the fiction experience by its creators. The appreciator can only actively participate in fourth wall breaks such as the Psycho Mantis one because they are first invited to do so by the videogame. In this last part of the chapter, I want to suggest that players can also *initiate* fourth wall breaks, regardless of these fourth wall breaks being anticipated by (the designer of) the game. This might seem like a strange suggestion. In literature or film, it would be simply impossible for appreciators to initiate fourth wall breaks. When a reader or spectator starts talking to a character within the fictional world, this would simply not entail a fourth wall break, because this interaction is not fictionally acknowledged within the work. When a spectator suddenly starts shouting at one of the characters in a play, the situation is somewhat different. Whether or not this disruption entails a fourth wall break then seems to depend on the reaction of the actors on stage. A fourth wall break only happens when these actors, still in the role of their character, acknowledge the audience member's utterance. In general, we could say that the occurrence of fourth wall breaks in literature, movies, and theatre always depends on the willingness of the fiction creator.

In videogames, however, appreciators can break the fourth wall in ways that are not anticipated by the videogame's designer.[2] This possibility for appreciator-initiated fourth wall breaks arises because players are not only actual people who know about the technology on which the game runs, but also take on the role of fictional characters within the fictional world. Players, as Conway already noted, take on a dual role as both audience member and performer (2010, p. 146). This brings about the weird situation in which a fictional character can undertake actions within a fictional gameworld that are motivated by knowledge of the artificiality of this world. Whenever the actual player reveals the technological apparatus underlying the fictional gameworld from their position within this world, this entails a fourth wall break, because this act will always be acknowledged by at least one fictional character: the player-character.

In *Uncharted: Drake's Fortune* (Naughty Dog, 2007), for example, players found out that, by aiming their gun in a certain way while standing next to a wall, they can eventually move through these walls. Doing this, they will often end up in unfinished locations in the gameworld that were never meant to be seen. Due to the obvious unfinished nature of these locations, the game reveals its own mediality as, in this case malfunctioning, software. But it only does so because the player, in the guise of a character in the gameworld, caused it to do so by misusing the game. And this player-character, by moving through these visually estranging spaces, fictionally acknowledges the existence of these locations. The player, actively and quite uncalled for by the game or its designers, thus broke the fourth wall.

Such player-initiated fourth wall breaks are not rare occurrences. Many videos on YouTube show players interacting with games in ways that were not intended by the game designers, and that show off ways to "break the fiction". Doctor Kill's "Let's Play Ocarina of Time Super Wrong" YouTube series (2015), for example, shows off ways in which players can misuse

[2] In their chapter in *Transgressions in Games and Play*, Mortensen and Navarro-Remesal already hint at the possibility of players breaking the fourth wall for other players in multiplayer games. They describe situations in which "the fourth wall is assaulted with direct addresses and in-jokes to break the unity of the ludo-fiction" (2019, p. 37).

the game *The Legend of Zelda: Ocarina of Time* (Nintendo EAD, 1998) to achieve comic or useful in-game effects. These videos can be interpreted as a behind-the-scenes documentary of the game, approached from within the gameworld itself. In the first video in the series, Doctor Kill, talking about the player-character Link from a first-person perspective, shows how a creative use of the in-game bomb items can be used to push Link out of the boundaries of the gameworld, showing parts of this world that should not at all be reachable in play. Aarseth, in this regard, talks about "transgressive play" or occurrences in which the player breaks out of the role that was designed for them to fulfil by the game's designers (2014, p. 132). Note that this description is already reminiscent of fourth-wall breaking characters who step out of their fictional role to acknowledge their own fictionality or (the audience in) the extra-fictional world.

The reason why this kind of transgressive play has not, up until now, been recognized as instances of breaking the fourth wall is because many game researchers took these situations as examples in which players do not interact with games as fiction, but rather as digital artefacts. Sageng, for example, writes that "the fictional pretence can peel entirely off if there is reason to think that it does not matter to the player" (Sageng, 2012, p. 230). If it is true that players who misuse games by approaching them purely as digital constructs do not at all interact with these games as fictions, then a fourth wall is never formed, and fourth wall breaks cannot take place.

However, I suggest to look at these situations from a different perspective. More specifically, I suggest not to treat speedrunners or otherwise mischievous players as altogether disregarding the game's fictional world, but rather as interacting with this world in a way that is heavily informed by, and self-referentially reveals, the fictionality and mediality of this world.[3] In other words: the fictional pretence does not peel off entirely in

[3] I do acknowledge that there are extreme cases in which game objects are interacted with while completely disregarding their fictional aspects. Examples of this are so-called game data mining or the reverse engineering of games, in which people look through a game's code (without even entering the game or its fictional world) or investigate how the game was made to find out what the use and functions of the game's objects are. Whenever a game's world is entered from a player perspective, however, there will always be an, albeit very minimal, fictional dimension.

the way they play, but this way of playing rather inherently involves fourth wall breaks. Another example that can clarify this are the speedrunning players of *Sekiro* (FromSoftware, 2019), who found ways to manipulate the game into rendering their avatar as if he is underwater while he is not. This enabled them to swiftly swim through the air from one location to another. The way players talk about this particular game exploit already reveals a certain fictional component to it: they talk of "airswimming" or being able to "fly", sometimes even referring to this power as a very special kind of "shinobi technique" (Whalen, 2019). It is true that, in this process of airswimming, the depicted fictional world loses coherency and is revealed to be a mere manipulable digital construct. But such incoherency and overt mediality are inherent to fourth wall breaks in any medium. When a movie character like Deadpool suddenly turns to the audience and starts interacting with them, for example, it is hard to make fictional sense of this (does he see the audience? Is he talking to a camera? Why cannot the other characters see us too?). Likewise, when Bugs Bunny gets into a fight with his illustrator, it is hard to truly care for him or the fictional world in which he lives, as both have just been revealed to be mere drawn constructs. In a similar way, the speedrunners of *Sekiro* reveal the fictional world of this game to be a mere piece of malfunctioning software. This, however, does not change the fact that the player still somehow interacts with this fictional world, and breaks its fourth wall in the process: namely, by taking on the role of a ninja who can swim through the air.

8 Conclusion

In this chapter, I have argued that the concept of fourth wall breaks, understood like breaks of the conventional relation between the real and the fictional world, is perfectly applicable to interactive fictional media like videogames. The use of digital technologies makes players imagine not only a fictional world, but also their own involvement within these worlds. Moreover, these technologies typically instruct players on how to interact with them by showing on-screen, self-referential information. As self-referentiality and self-involvement of the appreciator are inherent to videogames as virtually mediated fictions, they do not as such entail fourth wall breaks. They can, however, be creatively used to break the fictional gameworld's fourth wall. More importantly, virtually mediated fictional worlds such as those of videogames allow for creative interactions through the fourth wall from both the fictional characters' and the actual player's perspective. Due to players taking on the dual role of an external observer of the fictional world and a character within this world, they gain the power to break the fourth wall themselves. While videogames have often been said to be the first fictional medium to which the concept of the fourth wall is not applicable, I argue that they might be the first fictional medium that allows their appreciators to break this wall.

Due to the scope of this chapter, I have not addressed every puzzle that is connected to this phenomenon of appreciator-initiated fourth wall breaks, which deserves further attention. First of all, there are obvious differences between character-initiated and player-initiated fourth wall breaks. When a character breaks the fourth wall, they do not do so due to having any knowledge about the actual world or their own fictionality. At most, they are scripted to fictionally have such knowledge. In fact, when a movie character is designed to break the fourth wall by addressing spectators of the movie, it will still do so even when the movie is being played in an empty room. When a player, in the guise of a player-character, breaks the fourth wall, on the other hand, this action is motivated by real knowledge about the game as an artefact. This has two interesting consequences. First of all, the fourth wall break is less unexpected for the player who initiates it based on their knowledge of the gameworld as an artefact, thus losing the feeling of surprise that is often considered to characterize such breaks. Yet, despite their expectedness, player-initiated fourth wall breaks are still characterized by a disruptiveness, simply because the way the fictional world is represented during unconventional play is often visually weird and overtly technologically mediated. Secondly, the appreciator-initiated fourth wall break seems more true to the descriptions of fourth wall breaks as instances in which fictional characters somehow prove to have knowledge about the world outside of their fiction. After all, the player-character's fourth wall breaking actions are truly motivated by knowledge about the extra-fictional world: knowledge this character has because of its relation to the player. Undoubtedly, more can be said about the repercussions this kind of character-appreciator relation has for the boundaries of fictional gameworlds, especially regarding the ways in which these boundaries can be transgressed.

Lastly, I have briefly mentioned how player-initiated fourth wall breaks can be regarded as a subcategory of transgressive play (cf. Aarseth, 2014). Games, Aarseth writes, are "machines that sometimes allow their players to do unexpected things" (2014, p. 132). Transgressive play is such unexpected behaviour, which he describes as "a symbolic gesture of rebellion against the tyranny of the game, a (perhaps illusory) way for the played subject to regain their sense of identity and uniqueness through the mechanisms of the game itself" (2014, p. 132). In the case of player-initiated fourth wall breaks, such acts of rebellion are specifically aimed at the game's fictional world. As such, they are a testament to the kind of freedom players, in contrast to appreciators of movies or literature, have when playing games. As virtually mediated works of fiction, the videogame experience offers players the freedom not only to interact with the fictional content of the game and to decide how events will play out within the fictional gameworld, but also to interact with and transgress the very boundaries of that fictional world itself.

References

Aarseth, E. (2007). Doors and Perception. Fiction vs. Simulation in Games. *Intermediality, 9*, 35–44.

Aarseth, E. (2014). I Fought the Law: Transgressive Play and the Implied Player. In *From Literature to Cultural Literacy* (pp. 180–188). Palgrave Macmillan.

Bateman, C. (2011). *Imaginary Games*. Zero Books.

BioWare. (2011). *Mass Effect 2*. PlayStation 3: Electronic Arts.

Brown, D. W. (2012). *The Suspension of Disbelief in Videogames*. PhD thesis. Brunel University School of Arts.

Chalmers, D. J. (2017). The Virtual and the Real. *Disputatio, 9*(46), 309–352.

Conway, S. (2010). A Circular Wall? Reformulating the Fourth Wall for Videogames. *Journal of Gaming & Virtual Worlds, 2*, 145–155.

Currie, G. (1995). *Image and Mind: Film, Philosophy and Cognitive Science*. Cambridge University Press.

Diderot, D. (1947) [1758]. On Dramatic Poetry. In *European Theories of the Drama* (B. H. Clark, Trans.). New York: Crown Publishers.
Doctor Kill. (2015). Let's Play Ocarina of Time Super Wrong. *Youtube.* https://www.youtube.com/playlist?list=PLctnNKgSZgBzyu07wv3oiiQLNAIxOwuRI
Fox, T. (2015). *Undertale.* PC: Toby Fox.
FromSoftware. (2019). *Sekiro: Shadows Die Twice.* PlayStation 4: Activision.
Gualeni, S. (2019). On the De-Familiarizing and Re-ontologizing Effects of Glitches and Glitch-alikes. *Proceedings of the 2019 DiGRA International Conference*, Kyoto.
Jørgensen, K. (2012). Between the Game System and the Fictional World: A Study of Computer Game Interfaces. *Games and Culture, 7*(2), 142–163.
Jørgensen, K. (2013). *Gameworld Interfaces.* MIT Press.
Klevjer, R. (2006). *What is the Avatar? Fiction and Embodiment in Avatar-based Singleplayer Computer Games.* PhD Thesis. University of Bergen.
Konami Computer Entertainment Japan. (1998). *Metal Gear Solid.* PlayStation, PC: Konami.
Kubiński, P. (2014). Immersion vs. Emersive Effects in Videogames. In D. Stobbart & M. Evans (Eds.), *Play, Theory, and Practice: Engaging with Videogames.* Inter-Disciplinary Press.
Lamarque, P. (1981). How Can We Fear and Pity Fictions? *The British Journal of Aesthetics, 21*(4), 291–304.
Matravers, D. (2014). *Fiction and Narrative.* Oxford University Press.
McDonnell, N., & Wildman, N. (2019). Virtual Reality: Digital or Fictional? *Disputatio, 11*(55), 371–397.
Moradi, I. (2004). *Glitch Aesthetics.* Bachelor Thesis. Huddersfield, UK: University of Huddersfield. Retrieved November 22, 2018, from http://www.oculasm.org/glitch/download/Glitch_dissertation_print_with_pics.pdf
Mortensen, T. E., & Navarro-Remesal, V. (2019). Asynchronous Transgressions: Suffering, Relief, and Invasions in Nintendo's Miiverse. In K. Jørgensen & F. Karlsen (Eds.), *In Transgression in Games and Play* (pp. 27–44). Press.
Murray, J. H. (2017). *Hamlet on the Holodeck: The Future of Narrative in Cyberspace.* The Free Press.
Naughty Dog. (2007). *Uncharted: Drake's Fortune.* PlayStation 3: Sony Computer Entertainment.
Naughty Dog. (2020). *The Last of Us Part II.* PlayStation 4: Sony Interactive Entertainment.

Neill, A. (1993). Fiction and the Emotions. *American Philosophical Quarterly, 30*(1), 1–13.
Nintendo EAD. (1998). *The Legend of Zelda: Ocarina of Time.* Nintendo 64: Nintendo.
Nintendo EAD. (2001). *Animal Crossing.* Nintendo 64: Nintendo.
Robson, J., & Meskin, A. (2012). Fiction and Fictional Worlds in Videogames. In J. R. Sageng, H. J. Fossheim, & T. M. Larsen (Eds.), *The Philosophy of Computer Games* (pp. 201–217). Springer.
Robson, J., & Meskin, A. (2016). Video Games as Self-Involving Interactive Fictions. *The Journal of Aesthetics and Art Criticism, 74*(2), 165–177.
Rocksteady Studios. (2009). *Batman: Arkham Asylum.* PlayStation 3: Eidos Interactive.
Sageng, J. R. (2012). In-Game Action. In J. R. Sageng, H. J. Fossheim, & T. M. Larsen (Eds.), *The Philosophy of Computer Games* (pp. 219–232). Springer.
Sega AM3. (1993). *Sonic the Hedgehog.* Sega System 32: Sega.
Square Enix. (2005). *Kingdom Hearts II.* PlayStation 2: Square Enix Product Development Division 1.
Tavinor, G. (2005). Videogames and Interactive Fiction. *Philosophy and Literature, 29*(1), 24–40.
Tavinor, G. (2009). *The Art of Videogames.* Wiley-Blackwell.
Van de Mosselaer, N. (2018a). How Can We Be Moved to Shoot Zombies? A Paradox of Fictional Emotions and Actions in Interactive Fiction. *Journal of Literary Theory, 12*(2), 279–299.
Van de Mosselaer, N. (2018b). Fictionally Flipping Tetrominoes? Defining the Fictionality of a Videogame Player's Actions. *Journal of the Philosophy of Games, 1*(1). https://journals.uio.no/JPG/article/view/6035
Walton, K. (1978a). Fearing Fictions. *The Journal of Philosophy, 75*(1), 5–27.
Walton, K. (1978b). How Remote are Fictional Worlds from the Real World? *The Journal of Aesthetics and Art Criticism, 37*(1), 11–23.
Walton, K. (1990). *Mimesis as make-believe: On the foundations of the representational arts.* Harvard University Press.
Waszkiewicz, A. (2020). Together They Are Twofold': Player-Avatar Relationship Beyond the Fourth Wall. *Journal of Games Criticism, 4*(1). http://gamescriticism.org/articles/waszkiewicz-4-1
Weise, M. (2008). Press the 'Action' Button, Snake! The Art of Self-Reference in Video Games. *Game Career Guide* 25. https://gamecareerguide.com/features/652/press_the_action_button_snake_.php?page=1
Whalen, A. (2019). 'Sekiro' Flying Guide: How to Swim Through the Air With 'Shadows Die Twice' Glitch. *Newsweek.* https://www.newsweek.com/sekiro-flying-guide-glitch-shadows-die-twice-game-1392259

Games, Artworks, and Hybrids

Manuel García-Carpintero

1 Introduction: Videogames and Artworks

Videogames are objects that, at least in some cases and *prima facie*, have a dual nature: they are games, and they may also be (better or worse) artworks, more specifically *fictional* artworks. Some authors contest this view. From a Suits-inspired perspective on games that I broadly share, Rough (2018a) argues that games cannot also be artworks. Here I will outline an account of games on the one hand, and fictional artworks on the other, that validates the intuitive possibility of artwork videogames in

Financial support for my work was provided by the DGI, Spanish Government, research project PID2020-119588GB-I00; by the award *ICREA Academia* for excellence in research, 2018, funded by the Generalitat de Catalunya. A version of the paper was presented at the conferences *2nd VLC Philosophy Workshop: Sharing Reasons*, Valencia. I thank the audience for comments and suggestions. Thanks to Aarón Álvarez, Marta Campdelacreu, Filippo Contesi, Josep Corbí, Mohan Matthen, Enrico Terrone, and two referees for this volume for their comments, and to Michael Maudsley for the grammatical revision.

M. García-Carpintero (✉)
University of Barcelona, Barcelona, Spain
e-mail: m.garciacarpintero@ub.edu

E. Terrone, V. Tripodi (eds.), *Being and Value in Technology*,
https://doi.org/10.1007/978-3-030-88793-3_9

the face of Rough's arguments. I will rely for this on proposals articulated by García-Carpintero (2019, 2021a, 2021). The argument that I'll offer will be abductive. Some support will be given here for the controversial assumptions on which I'll rely in that they help answering Rough's concerns. Independent support has been provided for them in earlier work and in the tradition to which it belongs—the tradition of thinking of artworks as resulting from communicative acts understood on the model of institutions defined by constitutive rules, like games, following writers like Wittgenstein, Austin, or Danto.

This is, in outline, a preview of the argument: both game-kinds and artwork-kinds are defined by *accepted constitutive rules*, as understood in García-Carpintero (2021). There are functional accounts that explain why their defining rules are *accepted* or *enforced*. They depend on the *intrinsic value of experiences* of different sorts: the *aesthetic* and the *lusory* experience. Games and artworks thus have different functions. But individual artefacts can be hybrids with several functions, as illustrated by religious, political, or erotic artworks, by architecture and design. This is crucial to understand fictions, which might make assertoric acts, without being thereby "patchworks" of fiction and non-fiction (García-Carpintero, 2019, 2021b).

I'll start by explaining in the next section how institutions defined by constitutive rules should be understood; the material here mostly summarizes the account in García-Carpintero (2021). I make three claims: (i) kinds defined by constitutive rules (like particular games, and particular representational acts) are *real* kinds, with *real* definitions; (ii) their defining conditions are *normative*; (iii) for them not to be merely *putative*, their rules must be *accepted*. In Sect. 3, I'll explain the commonalities and differences between games and artworks, on my assumptions. Finally, in Sect. 4 I'll address Rough's arguments. Videogames as such will drop from the picture until that final section; I must beg readers for their patience. As my discussion of Rough will show, I need both a sufficiently elaborated account of games, and then one of artworks, for me to articulate a compelling enough response to his arguments.

2 Games as Kinds Defined by Constitutive Rules

In this section, I'll contrast two accounts of the distinction between constitutive and regulative rules, and I'll offer reasons to defend the one I'll need for my reply to Rough. On Searle's (1969, p. 34) intuitive characterization of the distinction, *regulative rules* "regulate a pre-existing activity, an activity whose existence is logically independent of the rules", while *constitutive rules* "create the possibility of new forms of behavior" (ibid., 35). As shown below, the distinction is only apt if made relative to a specific practice: (1) below is meant to illustrate regulative rules, (2) constitutive rules, for the same actual game, chess. I use imperative versions to critically address Searle's syntactical characterization of constitutive rules; but also the more malleable declarative modal formulation I take them to illocutionarily entail:[1]

(1) Control the center in the opening!
∴ One must (control the center in the opening)
(2) Castle only if the castling king is not in check!
∴ One must (castle only if the castling king is not in check)

I will use 'formalism' to refer to "the view that games (and sports to the extent that sports are games of physical skill) can be defined primarily by reference to the idea of constitutive rules and goals or obstacles designated by the rules themselves, which are unintelligible apart from them",

[1] I'll use 'rule' and 'norm' interchangeably. As I use these terms, rules are just what "*flat-out*" directives—typically conditional ones: do Q! (if P)—signify, to echo Williamson's (1996, p. 246) usage for the case of assertion: the specific force that is indicated in default, literal uses of imperative phrases. "Flat-out" directives are those uttered by someone presumed to have the required authority to make a command, as when I say 'Eat!' to my daughter, in contrast with 'Take it!' said to give advice, or 'Keep well!'. Kaufmann (2012) argues that a directive that *p* is synonymous with an assertion that the addressee should/must make it the case that *p*. On standard Kratzerian contextualist modal semantics, the obligation is conditional on implicit or explicit features that select accessible worlds and order them. This is implausible (Charlow, 2018, p. 82; Roberts, 2018, p. 331); but I do take directives to entail on illocutionary grounds Kaufman's assertoric counterparts (Charlow, 2014), and I'll avail myself of this throughout the paper.

Simon (2000, p. 2).[2] Formalism is the view that some rules of games are constitutive and not merely regulative, that is, that they provide the essence, nature or "real" definition of such kinds, thereby "creating their possibility". Formalism thus contends that practices are "natural kinds", in the contemporary sense that they have "real definitions".[3] Here 'natural' is not used in opposition to 'social', but rather to refer to properties and kinds in Lewis's (1983) "sparse" (as opposed to "abundant") sense. In fact, the account I'll assume for games and for illocutionary types like fiction and assertion takes them to be "social constructs", grounded on social rules.[4] "Natural" properties and kinds are those that play substantive explanatory roles, and hence have a "hidden nature" which only reveals itself after theorizing.

On a standard interpretation, anti-formalism just rejects that games have a rule-constituted essence. This is typically predicated on a Wittgensteinian, anti-definitional take on games that (Suits, 1978, p. 21) justly disparaged.[5] Thus understood, anti-formalism is not a compelling view. Kinds might not have definitions in terms of intrinsic properties, but they may still have perfectly fitting and explanatory relational ones, like those that have been claimed for biological and social kinds, or the one that defines an institution like *Analytic Philosophy* (García-Carpintero, 2011, 2021b). A more plausible version of anti-formalism doesn't deny that games and speech acts have definitions, but claims instead that, as in the case of 'jade', the ones they have are not deeply explanatory. Perhaps

---

[2] Similar characterizations can be found in the literature: according to formalism, "games are a product of their constitutive rules … these rules jointly create and define the game" (Kretchmar, 2015, p. 11); "the essential nature of a game is its rule-set and … proper play involves obeying the rules" (Nguyen, 2017, p. 9).

[3] Recent discussion of these issues mostly stems from the deservedly influential work of Kit Fine; cf. Correia (2017) for a good account and further references.

[4] They are "Platonic" essences, in the (fitting in this context, see below, fn. 10) terminology of Newman & Knobe (2019), as opposed to "causal" essences like that of water. I thus assume the "generalized essentialism" they argue for. It is controversial that artefacts (as opposed to the superordinate category *artefact* itself) have explanatory natures, but I am far from being idiosyncratic in thinking that they do have them (Evnine, 2016; Preston, 2020).

[5] As I make clear below, Sect. 3, unlike Suits I only aim to vindicate formalism for species-like "lower taxa"—in our context, specific games or specific speech acts. When it comes to genus-like "higher taxa" like *game* or *art* in general, or assertion understood as a genus encompassing guesses, conjectures, putting forward propositions for their consideration, and so on, the sort of definition I favor is closer to a nominal, non-explanatory one than a real definition.

they only have *cluster* definitions, or definitions in terms of exemplars or prototypes that just deploy their manifest, superficial, or criterial properties.

Constitutive rules, Searle says, "often have the form: X counts as Y in context C" (35); in a more recent piece he still sticks to this grammatical elucidation: "[i]t is [...] natural to think of these rules as having characteristically different syntactic forms. The regulative typically takes the form of an imperative, "Drive on the right!", for example; the constitutive rules, not so obviously but I think nonetheless apparently, take the form "X counts as Y" or "X counts as Y in context C", Searle (2018, p. 52). (2) thus becomes (3):

(3) *x counts as* castling in chess only if the castling king is not in check when *x* is performed

This strongly suggests the distinctive tenet of a variety of formalism I'll call "descriptive formalism", F_D, to wit, that the only 'must' in constitutive rules is not really prescriptive, but definitional; it is that of logical or conceptual necessity. Kinds defined by constitutive rules don't differ on this view from 'water', on the standard Kripke-Putnam line. We thus obtain the following application to our case of F_S:

(4) *x* is$_{df}$ castling in chess only if the castling king is not in check when *x* is performed

This is of course only a partial definition; other conditions should be added to obtain a full one: see rule 3.8 in the *FIDE Laws of Chess*.[6] Midgley (1959) and Suits (1967) explicitly embrace F_D; Guala (2016, pp. 58–59) follows a long trend of interpreting formalism that way. The most distinctive consequence of F_D is this: "To break a constitutive rule is to fail ... to play the game at all", Suits (1978, p. 52). Current

[6] As a matter of fact, as Searle (1969, p. 36) points out, only the whole set of rules will do. A standard way for formalists embracing F_D to react to the counterexamples discussed below is to restrict the claim of constitutivity to a designated set of rules, as it were the "important" or distinctive ones. García-Carpintero (2021) offers reasons to reject this move. See also fn. 13 below and related material in the main text.

supporters concur: "It is a rule of chess that a pawn moves a certain way, and one is not playing chess if one is not acting in accord with this rule", Jankovic (2018, p. 376, fn. 11); "rules are constitutive of the game and to not obey them is to not play the game", Rough (2018a, p. 7). According to F_D, when the rule is (intentionally) broken, the activity doesn't occur—the performance is an Austinian *misfire*, not an *abuse*.

Williamson (1996) offers an account of assertion in terms of constitutive rules, taking games as models. He hypothesizes (for abductive inquiry) that a single CR-shaped norm defines it, in which C stands for a suitable property of the expressed proposition:

(CR) One must (assert *p* only if *p* has C)

In explicitly drawn contrast to F_D, Williamson insists that the 'must' in (CR) is genuinely prescriptive: "Constitutive rules do not lay down necessary conditions for performing the constituted act. When one breaks a rule of a game, one does not thereby cease to be playing that game" (1996, p. 240). Assuming F_D, philosophers hence object that this view does not really assimilate speech acts to games (Johnson, 2018, p. 55); norms like (CR) don't have the proper Searlian form to be constitutive norms (Maitra, 2011, p. 282).[7] In my experience, Maitra's is the crucial consideration that moves defenders of F_D: if a norm is expressed in the imperative, that already makes it "regulative"; for a rule to be constitutive, it must be in the "count as" form.

However, as has been repeatedly pointed out in the literature, this syntactical characterization of the distinction is misguided. It is inspired by the right view that constitutive rules provide real definitions for their rule-constituted kinds; but the way it aims to capture this by means of a purely formal criterion is misguided. On the one hand, as Lewis (1979, p. 344) indicates and (2) illustrates, straightforward imperatives might well be constitutive rules in the intended sense (Ransdell, 1971, pp. 390–392; Glüer & Pagin, 1999, p. 220; Morin, 2011, p. 190). On the other, little ingenuity is required to frame any imperative in explicit

---

[7] Hindriks (2009, p. 272), Guala and Hindriks (2015, pp. 188–189), Guala (2016, pp. 58–59), and Reiland (2020, p. 7, fn. 9) all assume the linguistic characterization of the distinction in their discussions.

definitional locutions (Black, 1962, pp. 123–124; D'Agostino, 1981, p. 12; Glüer & Pagin, 1999, p. 217; Goldberg, 2015, p. 22):[8]

(5) *x counts as* an assertion of *p* (if and) only if: it is wrong (to perform *x* if *p* lacks C)/it is impermissible (to perform *x* if *p* lacks C)/one must (perform *x* only if *p* has C)
(6) *x* is_{df} an assertion of *p* (if and) only if: it is wrong (to perform *x* if *p* lacks C)/it is impermissible (to perform *x* if *p* lacks C)/one must (perform *x* only if *p* has C)

Williamson thus assumes a different variety of formalism, which I'll call "normative formalism", F_N. Rules of the F_N kind only differ from those of the F_D sort in having truly normative, deontic conditions in their *definiens*. They thus allow for Austinian *abuses*, in addition to *misfires*; or better *require* them, as Williamson points out: "The rule unconditionally forbids this combination: one asserts *p* when *p* lacks C. The combination is possible, otherwise it would be pointless to forbid it" (ibid., 254). But this is consistent with having the status of truly providing the constitutive nature of kinds, and hence with their character of constitutive norms, "creating the possibility of" new behavior. This is thus still *formalism*: it espouses the distinction between constitutive and regulative norms, and grants that the former provide the nature or essence of distinctive kinds.[9] It is the version disparagingly qualified as *Platonist* by the earlier critics who pointed out the obvious failures of F_D (D'Agostino, 1981, p. 9; Morgan, 1987, p. 2), because it primarily characterizes *an ideal* for games—what a *good* game should be. Games are *norm kinds* in

[8] This shows that the constitutive/regulative distinction can only be made relative to particular games. Searle begrudgingly admits as much: casting the rule in the "*X* counts as *Y* in context *C*" mold "is not intended as a formal criterion for distinguishing constitutive and regulative rules. Any regulative rule could be twisted into this form, e.g., 'Non-wearing of ties at dinner counts as wrong officer behavior'." But then he goes on to argue that, with some qualifications, the formal criterion would do; cf. García-Carpintero (2021) for critical discussion.

[9] Formalism as understood here is thus an ontological view. Just to be clear, I am not suggesting that Williamson is committed to its details; perhaps he doesn't understand "definition in terms of constitutive rules" in the way I am assuming. In fact, in a quotation I provided above he says that "[c]onstitutive rules do not lay down necessary conditions for performing the constituted act"; but in my view they do, if the defining conditions properly include normative vocabulary. Thanks to Samuele Chilovi and Matthew Kramer.

Wolterstorff's (1980, p. 56) sense, kinds "such that it is possible for them to have properly formed and also possible for them to have improperly formed examples".[10] On the F_N variety of formalism, the partial definition for castling in (4) should be replaced by (7):

(7) *x* is_{df} castling in chess only if it is not allowed if the castling king is under check

What are those "obvious" failures of F_D? It has the unacceptable consequence that cheaters, tactical breakers of rules, or inept players are not playing the games (D'Agostino, 1981, p. 9; Lehman, 1981, p. 42; Williamson, 1996, p. 240; Rescorla, 2009, p. 101; Kreider, 2011, p. 56; Ridge, 2017, p. 7; Fricker, 2017, p. 407; Consalvo, 2014) discuss cheating in videogames). Lehman mentions the case of a very prominent baseball player, the pitcher Gaylord Perry, who is in the baseball Hall of Fame even though it is generally accepted that throughout his career he threw *spitballs*—pitches that are illegal according to the rules of baseball, thrown after applying some foreign substance to the ball, causing it to do unpredictable things. The F_D account would entail that Perry never pitched in the relevant games, and nor, in consequence, did his teammates. Supporters of F_D have had recourse to different strategies to confront the problem; I just refer the reader to García-Carpintero (2021) for reasons to seriously doubt that they succeed.[11]

I have now presented the two first features of the view on rule-constituted kinds I rely on which I listed at the end of the previous section; I'll move now to the third, that for them not to be merely *putative*, they must be *accepted*. Rawls (1955, 26 fn.) distinguishes between *blueprint* kinds defined by constitutive rules, and those *that are in force* (cf. Glüer & Pagin, 1999, p. 221). There must be some account of why the rule constituted kind has come to be *in force* or

[10] It is also in this sense, I take it, that Newman and Knobe (2019) talk of "Platonic essences" in their defense of the generalized essentialism I am assuming here, see fn. 4 above.

[11] Suits (1978, p. 52) suggests a form of "legal positivism/realism", claiming that a move in a game is illegal if and only if a referee decrees it so. But this is a poorly motivated form of conceptual engineering. It leaves out games played without a referee (Kaluziński, 2018, p. 119). And it assumes a questionable view of what referees do, as Russell (1999) points out; referees also make mistakes, as shown in that their decisions are sometimes overruled.

accepted. Many kinds defined by constitutive norms are not in force, and hence are not really normative: they don't give anybody a reason to act.[12] Think of variations we can concoct on the rules that define actual games, and the "possible" games they define (I use scare quotes because, as abstract kinds, they are actual).[13] They define *putative*, candidate kinds; but, if they are not in force, nobody is really obligated by them. They don't give anybody a reason to act, and, in that way, they are not really normative. What would make such putative rule-constituted kinds "actual" in a given community, practices whose constitutive norms players are bound by?

What this requires is the "collective acceptance" that Searle (1995) discusses in more recent work, or Hart's (1961, p. 51–61) "rules of recognition". This might come to a convention or agreement; more in general, it is realized by social norms (Bicchieri, 2006; Brennan et al., 2013 ch.1, Southwood, 2019). There must be some related teleology, some "coordination problems" or goals more in general (Lewis, 1969) that are to be achieved by enforcing the norms (Graham, 2019).[14] For the case of games, this has been discussed under the topic of the game "ethos", "the purpose of the activity, which can guide both the selection of rules and behavior outside the rules" (Nguyen, 2017, p. 11)—which anti-formalists of different denominations (D'Agostino, 1981; Lehman, 1981; Morgan, 1987, §5, 2004) invoke in offering an alternative account, and formalists upholding F_D have also aimed to accommodate (Kretchmar, 2001, 2015). My view on this is the topic of the next section, in which I'll outline my

[12] Cf. Broome (2013, ch. 2) for a sharp discussion of the current Scanlon-inspired take on the "core" normative notions that I am assuming here. Franssen (2009) provides elaborate analyses of normative talk concerning artefacts in this framework.

[13] Cf. McPherson (2011, p. 232) on *chess* vs. *schmess*, and Kaplan (2021, fn. 10).

[14] Fricker (2017, p. 388) provides a detailed explanation; she shows that both a convention or a social or a moral norm, and a constitutive rule, might well underwrite the very same actually instantiated practice. In Epstein's (2015, chs. 6–7) ideology, constitutive rules provide real definitions for the kinds they individuate, and they must be grounded (Correia, 2017); in his terms, there must be corresponding "framing principles". Social norms provide what Epstein calls *anchors* for kinds that are in force. The external "norms" that explain why practices defined by constitutive norms are in force need not be more than regularities in the behavior of the relevant community, sanctioned by rewards and punishment; this is in fact the way both Lewis and Bicchieri think of them, cf. Hédoin (2015), Guala (2016, chs. 5–6), but see Brennan et al. (2013, ch. 2–4) and Southwood (2019) for a non-reductive alternative that I am sympathetic to.

account of how the constitutive rules of games are accepted in parallel to the account of how those that I take to define specific artworks are.

3 The *Lusory* and the *Aesthetic* Attitudes

To establish a first basis for my response to Rough's argument, in the previous section I have outlined an account of constitutive rules, and I have agreed with Suits that particular games, like species in biology ("lower taxa", Ereshefsky, 2017), have *real*, substantive, explanatory definitions—which, in the case of games are provided in part by their constitutive rules along formalist lines. The same applies to particular artworks, as I'll suggest momentarily. But unlike Suits I doubt that "higher taxa" (the genus to which they belong, respectively *game* and *artwork*) have more than non-explanatory *nominal* essences, given by their criterial features. The second main pillar for my response to Rough in the next section will be a functional account of the general category *game*, which I'll model on a well-known and well-developed related account for *artwork*. Most of the section will outline this familiar (even if controversial) functional account of *artwork*, referring the reader to the relevant literature for details.

Suits doesn't define *being a game* but *playing a game*: "to play a game is to attempt to achieve a specific state of affairs [*prelusory goal*], using only means permitted by rules [*lusory means*], where the rules prohibit use of more efficient in favor of less efficient means [*constitutive rules*], and where the rules are accepted just because they make possible such activity [*lusory attitude*]" (Suits, 1978, pp. 54–55). In short, he has it, playing a game is a "voluntary attempt to overcome unnecessary obstacles" (55). As Russell (2004, p. 146) nicely puts this core idea, "games create opportunities for developing certain human excellences by presenting obstacles that must be mastered and overcome in order to achieve the goal set by the game".

We cannot extract from this an acceptable, real definition of 'game'. For his purposes, Suits needed the "prelusory goal" of games to be given independently of the rules; something like *putting the ball inside the basket* in basketball. But he himself (Suits, 1978, p. 58) admits that in fact the goals of games (checkmate in chess) can only be properly defined inside the "institutions" that those games constitute.[15] Granting that, it is not clear that many of the things we should count as games have such goals—say, *singing games* like *ring-a-ring o' roses*, or *make-believe games* like *cops and robbers. Games of luck* like *rock, paper, scissors* strain in their turn the notion that games "create opportunities for developing certain human excellences". Suits deals with these issues by excluding some of these practices from the category of games, and by diluting the notions of "prelusory goal" and "obstacles" to include the others; but as a result, the account appears to overgenerate (Geras, 2009; Ellis, 2011; Berman, 2013; Ridge, 2019).

As I indicated in Sect. 2, it would be wrong to conclude from this that 'game' doesn't have a definition; what it lacks is a real, explanatory definition. Putnam himself held that *being* H_2O provides the real essence of the kind designated by 'water' only in the "predominant sense" of the term (Putnam, 1975, p. 239). He thereby assumed that it has other "senses", which can be in operation as a function of "context". One is that in which it means the same as "transparent, odorless, tasteless liquid filling rivers, lakes and oceans" (ibid.). This provides a much less explanatory (and extensionally different) kind, given by a cluster of the criterial features we use in applying the term, a "nominal essence". This polysemy that Putnam assumed has been amply confirmed in recent research (Tobia et al., 2020).

In fact, the polysemy of kind-terms goes beyond the two ends of the spectrum—real, explanatory, induction-supporting essences provided by intrinsic features versus non-explanatory nominal essences given by a

[15] Rough (2018a, 6 fn.) tries to salvage a notion of *prelusory goal* "separable" from the rules: "The prelusory goal may not always, however, be separable from the institution of the game itself. For instance, checkmate is not a mere configuration of objects in space, but requires the conventions that dictate what moves the pieces can make and what contexts count as checkmate. What makes the prelusory goal separable is that one can still achieve the prelusory goal without following all the rules of the game. Chess serves as an excellent example of this, as learning the game often involves setting up the board, or looking at illustrations of it, in positions that do not start at the beginning, a clear abrogation of the rules, but still follow the rules from there as a way to teach certain principles". This questionably assumes F_D—which Rough indeed holds, as a quote above shows. The whole set of chess rules is needed to define *being a checkmate*; but they are to be understood in accordance with F_N. Rough's cases don't pose a problem; the rules are indeed violated, not because of ineptitude or cheating, but for other good reasons; cf. also fn. 18.

cluster of stereotypical features, or similarity to a prototype or exemplar exhibiting them—that Putnam's view illustrates for the case of 'water'. There is, in addition, the more general type/token polysemy we of course have also to countenance here; I'll come back to it below for our cases. Games afford a fourth type, as I'll now show.

It is by now a standard view that biological species cannot be defined by intrinsic features (say, a genome), as Kripke and Putnam assumed; but there are relational alternatives that many researchers accept (Ereshefsky, 2017; Bird & Tobin, 2018). There are, among other models, Boyd's (1999) *Homeostatic Property Cluster* view, evolutionary-lineage views like cladism, and a few others.[16] I'll assume that these relational views may also capture real, explanatory, induction-supporting kinds.[17] Now, game-kinds as understood so far here are defined by all their constitutive rules. However, we (perhaps even more saliently) take, say, soccer before and after the introduction of VAR (which slightly changed the rules, and hence, strictly speaking on the present view, the game) as "the same game". We also count as soccer what kids play in the street, without proper goals, proper equipment or the proper number of players.[18] Now, kinds defined by constitutive norms that are thereby in force have a history. This allows for a further polysemy, on which kind-terms designate the related historical kind, continuous across relatively small changes in the defining constitutive rules—for soccer, perhaps including a stage when it was brought to England by Germanic settlers in the fifth to seventh centuries and continuing till the VAR-improved game of today (García-Carpintero, 2021; Ridge, 2020).[19] While in the case of biological species, the historical kind would perhaps constitute Putnam's

[16] Martínez (2020) offers a compelling revision of the HPC account.

[17] For good examples of how arguments for this may go, well-informed by current science, cf. Franklin-Hall (2021) and Khalidi (2021) on reasons why the sexes male vs. female (in general, and more clearly in particular species) are real kinds of such a historical sort.

[18] The latter case could also be treated along the lines suggested in fn. 13. As a matter of fact, the two accounts are complementary: the cases that Rough puts forward are chess in the historical sense, and they are also chess as defined by the current FIDE rules when this is understood in accordance with F_N; the rules are not being fully obeyed, for good reasons.

[19] Changes are "small" relative to the preservation of the distinctive "lusory means" that the game affords, allowing for distinctive lusory experiences vis-à-vis those afforded by other historically defined good games, kept in existence because of that. See below for elaboration.

"predominant sense", on the view advocated here it would be secondary for rule-constituted artefacts.[20]

In the case of 'artwork', cluster definitions closer to the "nominal essence" end of the spectrum have some defenders, cf. Longworth and Scarantino (2010), and similar proposals can be made for 'game', in a Wittgensteinian spirit free from definitional animadversion. I myself, however, support functional accounts for both 'game' and 'artwork', which capture better in my view their nature, and the open-texture character of the concepts. I nonetheless take the functional accounts to place them also close to the "nominal essence" end of the spectrum, for reasons I'll give below. The functional accounts I'll suggest rely on well-known proposals for artworks based on the notion of an *aesthetic experience*. They aim to capture what I take to be the most suggestive aspect of Suits' views—his appeal to the *lusory attitude*, "the acceptance of constitutive rules just so the activity made possible by it can occur" (1978, p. 54). I'll present first the structurally identical nominal definition I like for 'artwork' in terms of the related notions of the *aesthetic attitude* and the *aesthetic experience*, which relies on a more familiar literature.

Although it has distinctive aspects, the view of artworks I want to promote is sufficiently close to other proposals in the literature in the functionalist camp (cf. Stecker, 2010, ch. 5) to allow for it to be presented without much elaboration or argument. Artworks are artefacts with the function of producing specific *aesthetic experiences*. These I understand as in the *non-minimalist* tradition on which they are essentially good enough, pleasant in an extended sense in which even first exposure to John Cage's *4'33"* or a white painting by Rauschenberg may afford one.[21] Many philosophers since Kant's *Kritik des Urteilskraft* have offered compelling, recognizable characterizations of such experiences (under several

[20] Polysemy cannot always be explained by assuming a *core* or *central* sense, from which the others can be explained, cf. Falkum and Vicente (2015). But in some cases, this is a good working hypothesis; cf. Tyler and Evans (2003) on prepositions.

[21] Cf., e.g., Lind (1992). To be fully candid, I have to confess that I would find it utterly impossible to understand—alien—the behavior of someone who would be prepared to incur the costs I am to enjoy a celebrated new production of Strauss' *Rosenkavalier*, or a once-in-a-lifetime Giorgione, Vermeer, or Caravaggio exhibition, for the sake of being exposed for a second time to Rauschenberg's painting, or Cage's piece—whether or not the accidental sounds made by the audience in the performance are part of it, cf. Dodd (2018).

labels), including Dewey (1958), Beardsley (1969), Scruton (1974), Lind (1992), Walton (1993), Iseminger (2006), Budd (2007), Gaut (2007), Stecker (2010), Goldman (2013), Levinson (2016), Matthen (2017), Gorodeisky (2019).

This is Stecker's (2010, pp. 52–53) definition, which for our purposes I endorse: *the experience derived from attending in a discriminating manner to forms, qualities or meaningful features of things, attending to these for their own sake or the sake of a payoff intrinsic to this very experience.* On non-minimal views, the aesthetic experience has positive hedonic value. This value is one of the "facilitating" kind, in contrast with the "relief" kind to which other pleasures belong (Matthen, 2017; cf. also Lind, 1980; Miller, 1998). It is certainly not "mere pleasure" (Gorodeisky, 2019); in my own estimate, particularly good items of the kind are afforded by Henry James' later novels, or Richard Strauss' operas, which certainly require a lot of effort for one to properly discern their rewarding features. On Lind's, Miller's and Matthen's accounts, what the relevant pleasure facilitates is the persistence in the effort required to engage an object with an *aesthetic attitude*. This is the attitude that, addressed to the proper object and in the proper enabling conditions, might be conducive to a fitting aesthetic experience—a satisfying "learning-like" response, Miller (1998). When the object of the attitude is an artwork, and hence defined by constitutive rules as I'll explain in a moment, Suits' characterization of the lusory attitude also fits the aesthetic attitude quite well, for it then is "the acceptance of constitutive rules just so the activity made possible by it can occur". In such a case, the activity in question is that of engaging the artwork in pursuit of the aesthetic experience it promises to provide—at least until one realizes that it will fail to live up to it.

Minimalist views about the aesthetic experience (Kivy, 2011; Carroll, 2012, 2016) deprive it of positive value. Their defenders are right that we also need the notions of an attitude which is specifically aesthetic and whose exercise doesn't guarantee any reward, what I am calling on Suits' inspiration *the aesthetic attitude*, which hence differs from the *aesthetic experience*.[22] This is because artworks may not be fit to produce aesthetic

[22] Of course, there is a long tradition of appealing to an aesthetic attitude in characterizing artworks, from Kant to Stolnitz and Scruton, cf. King (2012). The Suits-inspired role it plays in the

experiences, no matter how competently the aesthetic attitude is addressed to them. (This doesn't mean that they in fact do not produce them; they can produce *unfitting* aesthetic experiences.) Artworks I take to have the function, and hence to be subject to the norm, of allowing (i.e., being normative reasons) for satisfying aesthetic experiences, when their enabling conditions are fulfilled:[23] one belongs in the class of subjects targeted by them, one is in the relevant age range, has the required expertise, has slept well, is not otherwise distracted, and so on and so forth. I am treating 'satisfying aesthetic experience' as the positive form of the comparative 'better aesthetic experience than', on the model of the standard semantic treatment of context-dependent gradable adjectives like 'tall' (cf., e.g., Kennedy, 2007; cf. Miller, 1998) on the "normal scale" of aesthetic appreciation). I am thinking of the aesthetic experience on the *fitting attitude* model that I think offers the best account for attitudes like emotions.[24] Emotions like *admiration* or *contempt* represent states of affairs; they might represent the same particular state of affairs—say, a violent act. What is distinctive about them is the conditions under which

present account makes it differ importantly from the way King defines it. It thus withstands the objections aptly discussed by Kemp (1999), taking up Kemp's demand that the discussion of its nature be placed "in the context of a suitably rich theoretical atmosphere" (ibid., p. 399).

[23] I am assuming the now standard Dancy-Raz take on *normative* and *explanatory* reasons; cf. Álvarez (2017). I think of functions along the etiological lines nicely articulated by Eaton (2020). On the view I am outlining, the etiology determining *proper function* and *proper conditions* for function fulfilment is afforded by the account of how the constitutive norms defining the artefact (the artwork) have come to be in place or accepted; cf. Thomasson (2014). As Eaton points out, this allows for artworks to be answerable to the relevant norms even when their makers lack the intentions to yield to them: individuals may be obligated by norms even when they lack such intentions, for different reasons. However, the normative reasons that I am assuming artworks should provide are meant, in prototypical cases, to be *owned by*, *possessed by*, or *available to* artwork producers and their audiences; I take it that a form of "moderate intentionalism" (Livingston, 2005) about proper interpretation (i.e., "Does Mean" interpretation, as opposed to "Could Mean" interpretation, Levinson, 1998) follows from this.

[24] Although I agree with Gorodeisky and Marcus (2018) that successful artworks provide normative reasons for an attitude, my view of the aesthetic experience is much closer to the one nicely articulated by Goffin (2019)—and the more general view of pleasant feelings developed by Aydede (2018). I sharply distinguish aesthetic experiences from the aesthetic judgments they may ground and justify; for the judgment can occur without the experience, as the experience can occur without the judgment. (My take on the related debate about aesthetic testimony is close to Franzén's (2018)). And, while I agree with Gorodeisky and Marcus (2018) and Gorodeisky (2019)—and Kant—that we intuitively take aesthetic experiences to make claims of necessity and universalizability, I think we have good theoretical reasons to resist them, and to embrace a form of (contextualist) relativism (Marques & García-Carpintero, 2014).

that state of affairs makes them *fitting*; whether it is a normative reason for them (Jacobson, 2011): contempt but not admiration might be the fitting response, or the other way around.

On the present ontological view of artworks, we thus need both psychological categories, *aesthetic attitude* and *aesthetic experience*, the latter being explanatorily fundamental. Other important related notions, like *aesthetic property* and *aesthetic judgment*, are also to be understood on its basis (Budd, 2007, pp. 46–47). As indicated at the outset, the resulting view is close to Danto's, cp. Stecker (2010, p. 108), Lind (1992, pp. 117–118). The fundamental ontological trait of artworks is that they are historically bound artefacts with a communicative purpose. They have a "text" distinguishable from the work that they are (Currie, 1991), as motivated by Borges' "Pierre Menard", or similar Danto cases. This "text" is the meaning-vehicle, constituted by aesthetic features that are targets for the conscious attention (sensuous in music or visual arts, cognitive in literature) that is a central focus of the aesthetic attitude (Lind, 1980). The work is additionally constituted by its meaning; this will include a content, but also force-like features, which can be assertoric, as in documentaries, an appeal to the imagination in fiction, or some other, expressive or conative. It is not so ontologically decisive whether the work is a repeatable (but temporally bound) universal (Thomasson, 1999), or a particular concrete object.

Let me explain how I think we should think of the meaning of artworks on the "language games" view as characterized in the previous section. As indicated there, Williamson (1996) offers a constitutive-rules account of the proper F_N sort for the specific kind he calls "flat-out assertion", which is according to him (ibid., 258) what we do by default when uttering sentences in the declarative mood. To interpret this, I (García-Carpintero, 2021a) appeal to the near-universal character of the three moods, declarative, imperative, interrogative; I assume that they are polysemic, and I take each of them to have a default use; these would be flat-out assertions, flat-out directives, and flat-out questions. A criterial feature of each of them would thus be the default meaning of each mood. Additional criterial features of flat-out assertions include their being used to provide information by being "taken at one's word", that is, based on one's presumed epistemic authority; for the imperative it would be to instruct based on one's authority, and for the interrogative to tell the audience to provide an answer on the basis of one's authority to that effect. Something like this is standardly assumed by linguists to be the functional role of the moods when they look for their paradigms in different languages (König & Siemund, 2007, pp. 282–284). Williamson goes on to offer a particular implementation of the "simple rule" model (CR) described in the previous section to characterize the nature of this specific speech act, a flat-out assertion with content *p*, (KR); some other proposals have been made, in particular the one I favor—the audience-involving rule (KPR), cf. Hinchman (2019):

(KR) One must ((assert *p*) only if one knows *p*)
(KPR) One must ((assert *p*) only if one's audience is put thereby in a position to know *p*).

As indicated above, a crucial aspect of a full defense of any of these views would have to include an account of why the relevant rule is accepted; on the model I have been suggesting, this would be a functional, etiological account, and the value that such an account will mention in this case will be the pooling of information through the transmission of knowledge via testimony (Graham, 2019). Such an explanation would establish that a practice whose essence is defined by (KPR) is a real, explanatory kind. For it would explain the criterial, manifest features of flat-out assertion mentioned above. The full explanation would involve a compositional semantics, to account for how the relevant content is determined; and it would have to include an explanation of how implementing a practice on which people in the relevant community are bound by the rule—so that they can be criticized if they make an assertion that they shouldn't have made, and so on—allows for the goals ascribed to the practice to be attained. This semantics for the relevant meaning-vehicles will have to be backed up by a metasemantics—an account of why the semantics is the proper one for them. If the constitutive rules view outlined here is correct, such metasemantics will rely on it, cf. (García-Carpintero, 2021a). But this doesn't detract from the explanatory significance of the Austinian account of assertion: the full explanatory package might have been provided instead by an account of a very different sort, a Gricean one, or a Chomskian one, say. If true, it would be illuminating to know that in default uses utterances of sentences in the declarative mood may allow for the transmission of knowledge along the lines of the explanation provided by the constitutive rules account.

Parallel points apply to artworks, in particular fictional ones. On García-Carpintero's (2013, 2019) development of Walton's views in the just outlined constitutive rules framework of fictional artworks, they are meant to furnish specific imaginative projects; and when it comes to their specific trait of being *fictional* artworks, they are constitutively liable to the following rule:

(FR) One must ((fiction-make *p*) only if one's target audience is put thereby in a position to enjoy a satisfying (good enough) imaginative project vis-à-vis *p*).

As above, this assumes that we can compare how satisfying the imaginative projects afforded by fictional artworks are for the intended audience (along multiple dimensions: to what extent we are caught up in the plot, how much we empathize with the characters or find them interesting, what imagery it prompts, how emotionally engaged we feel …), as part of a promised full aesthetic experience. It further assumes that the context in which the norm is in force provides an (indeterminate) cut-off point for the positive form mentioned in (FR).

It is essential that the constitutive norms for these communicative acts, flat-out assertion, and artwork fiction-making—(KPR) and (FR) respectively, I assume—be understood in accordance with F_N: clearly, a flat-out assertion may be a lie that cannot put anybody in a position to acquire knowledge of its content, and the imaginative project proposed by a fictional artwork may well be utterly uninteresting to the intended audience. Such failed acts do not afford normative reasons for belief or imagining vis-à-vis their (sufficiently determinate) contents, respectively; in that sense, they don't really prescribe beliefs, or imaginings. The metasemantics will establish that they are liable to such norms, relative to sufficiently well determined contents. They "prescribe" belief and imaginings with such contents only in the sense that—to focus on the fiction case—one should so imagine when deploying the aesthetic attitude in the proper circumstances, to establish whether the works fit the norm.[25]

The proposal thus characterizes specific assertoric speech acts and specific artworks as real kinds with explanatory normative essences; but it hardly goes beyond offering a nominal essence for higher-order kinds. Consider the kind *artwork*. What the outlined functional account in fact does is to appeal to our previous capacity to identify prototypical exemplars of the kind, by relying on our having enjoyed aesthetic experiences: "If there is one thing we know about aesthetic objects right off, it is that they are interesting in some way", Lind (1980). I have tacitly assumed the Louis Armstrong dictum, "if you have to ask … you'll never know" to apply to the aesthetic experience. I hope some illumination is provided by the functional account, but it hardly goes beyond highlighting the relevant features of the criterial, stereotypical traits.

All this applies, *mutatis mutandis*, to specific games, and to the kind *game*. I said that what I like in Suits' account is his characterization of the *lusory attitude*, "the acceptance of constitutive rules just so the activity made possible by it can occur", but we should ask, why do we indulge it? Why is it that we are happy to perform the "voluntary attempt to overcome unnecessary obstacles" that it demands? The reason is that good games, like good artworks, afford players similarly pleasant experiences, which similarly facilitate the effort that Suits describes—the effort of seeking their constitutive goals while abiding by their constitutive rules (Ridge, 2019, p. 78). These would be *lusory experiences*; Suits' *lusory attitude* is the attitude of actively seeking them—actively seeking the pleasure the game promises to afford, appraising it in doing so.[26] This applies to the sort of games that Suits took as prototypical, like chess or soccer;

---

[25] This paragraph was motivated by an illuminating exchange with Kendall Walton. The point in the text is compatible with the metasemantics allowing for some "value-maximizing" elements in *Does Mean* interpretation; for instance, to allow for a flat-out assertion of 'the vote was unanimous' to *say* that the vote was anonymous when the malapropism is sufficiently manifest (Unnsteinsson, 2017); or to allow the application of (FR) to Fritz Lang's 1937 *You Only Live Once* to be made relative to George Wilson's (1986, ch. 2) non-obvious (but to a very great extent upheld by clues in the film) interpretation, as opposed to the sentimental standard one.

[26] In a similar vein, Juul (2005, p. 226) states that a "game has been designed to be entertaining when one pursues the goal" (thanks to Aarón Álvarez and Alfonso García for the reference). This goal is what rules like the "12 Tenets of Board Game Design for Stonemaier Games", https://stonemaiergames.com/about/mission-statement/, transparently aim to secure.

but it applies also to the other sorts mentioned above, including those that he dismissed like singing games. Lusory and aesthetic experiences share crucial features: both are autotelic, sought for their own sake; both may fully occupy attention; both facilitate difficult activities. They differ in that these activities are different. The latter assists the activity of contemplating the "text" of an *artwork* while grasping its meaning. The former, the activity of playing a game, being in so doing thereby beholden to its specific constitutive rules.[27] On the present view, in sum, games are artefacts defined by constitutive rules, whose function is to produce in appropriate judges/players a satisfying enough lusory experience—one above a certain threshold.[28]

I have now developed the two pillars for my response to Rough—an account of kinds defined by constitutive rules that applies to games and language games, and a functional account of the general categories *game* and *artwork* predicated on the attitudes and the experiences they are meant to mobilize. We can move now to the issue whether games can be artworks.

---

[27] A referee asked how to locate my discussion in the long debate in the literature on games on the distinction traced in English (but not in Spanish and other languages) between 'game' and 'play'. It is generally granted that the latter has a wider use than the one at stake here (cf. Aarseth, 2014; Clément, 2014). For my purposes, 'game' refers to the abstract object defined by constitutive rules, 'play' to the activity of engaging it with the lusory attitude.

[28] I'd like to note here the most significant difference between my views on fiction and Kendall Walton's, in many ways the closest to my views and certainly the one that has been most influential in developing them. As elaborated in Chap. 2 of Walton (1990) and the earlier works that it crystallizes, Walton famously assimilates fictional artworks to make-believe games, and other categories including daydreams, and even dreams; Corbí (2020) identifies social rituals, and Stock (2021, ch. 6) legal fictions, and legal-fiction-like gender classifications, as other examples in the category that Walton calls "representations"—items that can be seen as having the prop-facilitated function of prescribing imaginings. Accordingly, Walton (2015a) takes sport games to be fictions. Now, I don't have any qualms in acknowledging Walton's ontological category, or its significance. However, as commentators have pointed out (Friend, 2008, pp. 152–154; Woodward, 2014, p. 825), Walton is dismissive of the narrower category that, say, Searle (1974–1975) or Currie (1990) were trying to account for—which is, I take it, that of fictional artworks. This cannot be explained in terms of the blanket notion of *prescriptions to imagine*, as Walton (2015b) now acknowledges; but as García-Carpintero (2019) shows, it can be along the lines outlined here. And it differs from games for the reasons given in the main text; as Walton (2015a, p. 80) himself points out, "even in the case of spectator sports like professional baseball games and track meets, no one arranges the events of the game to best advantage for appreciation—at least no one is supposed to. The participants play to win, not to put on a good show".

4 The Possibility of Hybrids: A Rejoinder to Rough

The views just outlined overlap significantly with Rough's (2018a), beyond our sympathy toward Suits' account of games. We agree that particular games and artworks are real kinds (ibid., p. 8). We agree that the general categories to which they belong differ: "Artworks and games are ontologically distinct things" (p. 2), even though they have something important in common: "Games and artworks both have among their constitutive parts prescriptions for how to appropriately engage and appreciate them" (p. 2). The difference lies for both of us in the kind of "engagement and appreciation" (p. 3) that items in each category demand; Rough assumes a functionalist view of artworks parallel to the one just articulated (pp. 8–11). But there are also important disagreements. Rough understands constitutive rules on the wrong model F_D. He fully endorses Suits' account of the kind *game*, which I have questioned. He mixes up the psychological and the normative. This will be the basis for my rejection of his arguments.

Rough summarizes his three main considerations thus: "(1) games and artworks cannot share the same constitutive goals, (2) artworks cannot employ the arbitrary inefficiency that is required of games, and (3) the required attitude that we take towards artworks undermines the required sufficiency of the attitude that we must take towards games" (p. 5). In developing these points, Rough appears to assume a hylomorphic metaphysics (Evnine, 2016; Fine, 2017), on which objects can share a "matter" (and hence spatiotemporal location) while still differing in their "form", and hence being different objects, like the notorious statue and piece of clay: "This incompatibility [between games and artworks] does not mean that a game, like most anything, could not be a constitutive element of an artwork, but a game cannot be identical with an artwork", p. 20. But Rough's claim doesn't follow from hylomorphism alone: the "form" of an object might combine that of an artwork, and that of a game. There are cameras; there are cellphones; and there are *hybrids*: artefacts that constitutively are both (Evnine, 2013, p. 415, fn.).

Rough's arguments for the first two objections to this natural proposal elaborate the point that artworks are "unable to have separable prelusory goals", and that there cannot be in their case "inefficient lusory means" (p. 20). Deploying the aesthetic attitude is a "technical" activity (p. 11), for we are supposed to use the most efficient means for our goal—proper appreciation of the work; the lusory attitude seeks instead enjoyment by using inefficient means to reach the goals set up by the game. This argument depends too much on Suits' general view of the genus-kind *game*, which we have found questionable (fn. 15 and text to which it is attached). But the main problem is that it confuses the psychological with the normative. Perhaps, as a matter of physical fact, it is impossible for the cellphone camera to take good pictures and simultaneously offer efficient communication; but it can be an excellent exemplar of both a camera and a cellphone nonetheless. Similarly, perhaps as a matter of psychological fact an agent can only enjoy at a time a videogame as a good game, or as a good artwork; perhaps the lusory and aesthetic attitudes "do not overlap" (17)—this should be a matter for empirical investigation, although, as far as I can tell first-personally, holding both together vis-à-vis one and the same object may well be psychologically feasible. But the object as such can still meet both standards perfectly well.

This confusion of the psychological and the normative starts already when Rough provides his own way of turning Suits' definition of *playing a game* into a definition of *game*. Rough says that games prescribe "that players engage the set of rules with the lusory attitude" (p. 6). This can at most be metaphorically true. Strictly speaking, games just prescribe for their constitutive rules to be obeyed. Players might play them fully competently without adopting the lusory attitude at all. Perhaps they are seasoned pros, who lost interest in the game long ago and are only motivated by money and fame. The fact that games assure (have the function of providing) lusory experiences and are thus to be engaged with the lusory attitude is only part of the explanation of why their constitutive rules are in force or accepted. The main problem with Rough's claims is that they confuse the constitutive rules that provide the essence of games, with the explanation—in part psychological—for why some of them are in force or accepted.

Rough's third critical point is that "the lusory attitude's sufficiency" is "undermined by the necessity of the artistic attitude" (2018a, p. 20). This doesn't conflate the psychological and the ontological (see also pp. 16, 17). But it simply begs the question to assume that the "lusory attitude" suffices for full appreciation of hybrid objects that are both games and artworks. "[G]ames prescribe the lusory attitude as attitudinally sufficient while artworks prescribe the artistic attitude as attitudinally necessary", p. 17. That is correct; but hybrid games-artworks do not "prescribe" either the lusory or the aesthetic attitude as attitudinally sufficient. The lusory attitude is only secure to suffice for full appreciation for objects that are (good enough) games, not for those that are both games and artworks.[29]

The points I have made in response to Rough apply equally to much more common cases of hybrid artworks. Documentaries are artworks (very good ones in some cases, e.g. Errol Morris' 1988 *The Thin Blue Line* or Joshua Oppenheimer's 2012 *The Art of Killing*) the force-component of whose meaning is assertoric; they may be just flat-out assertions meaning to provide testimony-based knowledge, or make informed guesses, or just put theories or propositions for our consideration. As Stacie Friend (2008, 2012) has been insisting, however, most fictional artworks are also meant to make assertoric acts; and the other way around: non-fiction artworks like the two documentaries I just mentioned also include, as essential components, fictional reenactments. Friend (2012) uses these facts to defend (I take it) a Wittgensteinian relational view on the *fiction* category, on which "prescriptions to imagine" are not essential features of fictions. Although I cannot develop this point here, the hybrid view of the relevant works allows us to resist her proposal (cf. García-Carpintero, 2021b).

In this chapter I have articulated parallel functionalist views of games and artworks, and I have relied on them to defend the possibility of

[29] My references in Sect. 2 acknowledge how close my views on games are to Mike Ridge's. Unsurprisingly, his response to Rough (Ridge, 2018) is also similar to mine. However, I grant more to Rough than he does. I grant Rough the hylomorphism, which Ridge objects to as leading to a "bloated ontology", p. 8—something that doesn't worry me in the least. More importantly, I think of both games and artworks as (in the primary sense of the terms for the relevant kinds) constituted by rules, along similar lines to Rough. This is why my own objections stand in the face of Rough's (2018b) brief dismissal of Ridge's objections.

hybrid artefacts that are both from Rough's (2018a) arguments. Given that I take the possibility—even the actuality—of such hybrids to be intuitively very plausible, my defense, if successful, offers a measure of abductive support for the controversial views on games and artworks on which I have relied.

References

Aarseth, E. (2014). Ontology. In M. J. P. Wolf & B. Perron (Eds.), *The Routledge Companion to Video Game Studies* (pp. 484–492). Routledge.

Álvarez, M. (2017): Reasons for Action: Justification, Motivation, Explanation. In E. N. Zalta (ed.), *The Stanford Encyclopedia of Philosophy* (Winter 2017 Edition). https://plato.stanford.edu/archives/win2017/entries/reasons-just-vs-expl/

Aydede, M. (2018). A Contemporary Account of Sensory Pleasure. In L. Shapiro (Ed.), *Pleasure: a History*. Oxford University Press.

Beardsley, M. (1969). Aesthetic Experience Regained. *The Journal of Aesthetics and Art Criticism, 28*, 3–11.

Berman, M. (2013). Sprints, Sports, and Suits. *Journal of the Philosophy of Sport, 40*, 163–176.

Bicchieri, C. (2006). *The Grammar of Society: The Nature and Dynamics of Social Norms*. CUP.

Bird, A., & Tobin, E. (2018). Natural Kinds. In E. N. Zalta (ed.), *The Stanford Encyclopedia of Philosophy* (Spring 2018 Edition). https://plato.stanford.edu/archives/spr2018/entries/natural-kinds/

Black, M. (1962). The Analysis of Rules. In *Models and Metaphors* (pp. 95–139). Cornell University Press.

Boyd, R. (1999). Homeostasis, Species, and Higher Taxa. In R. Wilson (Ed.), *Species: New Interdisciplinary Essays* (pp. 141–185). MIT Press.

Brennan, G., Eriksson, L., Goodin, R., & Southwood, N. (2013). *Explaining Norms*. Oxford University Press.

Broome, J. (2013). *Rationality Through Reasoning*. Wiley Blackwell.

Budd, M. (2007). Aesthetic Essence. In R. Shusterman & A. Tomlin (Eds.), *The Value of Aesthetic Experience* (pp. 17–30). Routledge.

Carroll, N. (2012). Recent Approaches to Aesthetic Experience. *The Journal of Aesthetics and Art Criticism, 70*(2), 165–177.

Carroll, N. (2016). Art Appreciation. *Journal of Aesthetic Education, 50*(4), 1–14.

Charlow, N. (2014). The Meaning of Imperatives. *Philosophy Compass, 9*(8), 540–555.
Charlow, N. (2018). Clause-Type, Force, and Normative Judgment in the Semantics of Imperatives. In D. Fogal, D. Harris, & M. Moss (Eds.), *New Work in Speech Acts* (pp. 67–98). OUP.
Clément, F. (2014). Players/Gamers. In M. J. P. Wolf & B. Perron (Eds.), *The Routledge Companion to Video Game Studies* (pp. 197–203). Routledge.
Consalvo, M. (2014). Cheating. In M. J. P. Wolf & B. Perron (Eds.), *The Routledge Companion to Video Game Studies* (pp. 150–157). Routledge.
Corbí, J. E. (2020). A Family Meal as Fiction. *Organon F, 27*(1), 82–105.
Correia, F. (2017). Real Definitions. *Philosophical Issues, 27*, 52–73.
Currie, G. (1990). *The Nature of Fiction*. Cambridge University Press.
Currie, G. (1991). Work and Text. *Mind, 100*, 325–340.
D'Agostino, F. (1981). The Ethos of Games. *Journal of the Philosophy of Sport, 8*, 7–18.
Dewey, J. (1958). *Art as Experience*. Capricon.
Dodd, J. (2018). What *4' 33"* Is. *Australasian Journal of Philosophy, 96*(4), 629–641.
Eaton, A. (2020). Artifacts and Their Functions. In I. Gaskell & S. A. Carter (Eds.), *Oxford Handbook of History and Material Culture*. Oxford University Press. https://doi.org/10.1093/oxfordhb/9780199341764.013.26
Ellis, J. (2011). On the Concept of a Game. *Philosophical Investigations, 34*, 381–391.
Epstein, B. (2015). *The Ant Trap*. OUP.
Ereshefsky, M. (2017): Species. In E. N. Zalta (ed.), *The Stanford Encyclopedia of Philosophy* (Fall 2017 Edition). https://plato.stanford.edu/archives/fall2017/entries/species/
Evnine, S. (2013). Ready-Mades: Ontology and Aesthetics. *British Journal of Aesthetics, 53*(4), 407–423.
Evnine, S. (2016). *Making Objects and Events*. Oxford University Press.
Falkum, I., & Vicente, A. (2015). Polysemy: Current Perspectives and Approaches. *Lingua, 157*, 1–16.
Fine, K. (2017). Form. *Journal of Philosophy, 114*, 509–535.
Franklin-Hall, L. (2021). The Animal Sexes as Explanatory Kinds. In S. Dasgupta, R. Dotan, & B. Weslake (Eds.), *Current Controversies in Philosophy of Science* (pp. 177–197). Routledge.

Franssen, M. (2009). Artefacts and Normativity. In A. Meijers (Ed.), *Philosophy of Technology and Engineering Sciences (Handbook of the Philosophy of Science, Volume 9)* (pp. 923–952). Elsevier.

Franzén, N. (2018). Aesthetic Evaluation & First-Hand Experience. *Australasian Journal of Philosophy, 96*(4), 669–682.

Fricker, E. (2017). Norms, Constitutive and Social, and Assertion. *American Philosophical Quarterly, 54*(4), 397–418.

Friend, S. (2008). Imagining Fact and Fiction. In K. Stock & K. Thomson-Jones (Eds.), *New Waves in Aesthetics* (pp. 150–169). Palgrave Macmillan.

Friend, S. (2012). Fiction as a Genre. *Proceedings of the Aristotelian Society, 92*, 179–208.

García-Carpintero, M. (2021). How to Understand Rule-Constituted Kinds. *Review of Philosophy and Psychology*.

García-Carpintero, M. (2021a). Metasemantics: A Normative Perspective (and the Case of Mood). In P. Stalmaszczyk (Ed.), *The Cambridge Handbook of the Philosophy of Language* (pp. 401–418). CUP.

García-Carpintero, M. (2021b). Documentaries and the Fiction/Nonfiction Divide. *Studies in Documentary Film*. https://doi.org/10.1080/17503280.2021.1923146

García-Carpintero, M. (2011). A Genealogical Notion. *Teorema, 30*, 43–52.

García-Carpintero, M. (2013). Norms of Fiction-Making. *British Journal of Aesthetics, 53*, 339–357.

García-Carpintero, M. (2019). Normative Fiction-Making and the World of the Fiction. *Journal of Aesthetics and Art Criticism, 77*(3), 267–279.

Gaut, B. (2007). *Art, Emotions and Ethics*. OUP.

Geras, N. (2009). Games and Meanings. In S. De Wijze et al. (Eds.), *Hillel Steiner and the Anatomy of Justice: Themes and Challenges*. Routledge.

Glüer, K., & Pagin, P. (1999). Rules of meaning and Practical Reasoning. *Synthese, 117*, 207–227.

Goffin, K. (2019). The Affective Experience of Aesthetic Properties. *Pacific Philosophical Quarterly, 100*, 283–300.

Goldberg, S. (2015). *Assertion*. OUP.

Goldman, A. (2013). The Broad View of Aesthetic Experience. *The Journal of Aesthetics and Art Criticism, 71*(4), 323–333.

Gorodeisky, K. (2019). The Authority of Pleasure. *Noûs* online first. https://doi.org/10.1111/nous.12310

Gorodeisky, K., & Marcus, E. (2018). Aesthetic Rationality. *Journal of Philosophy, 115*(3), 113–140.

Graham, P. J. (2019). The Function of Assertion and Social Norms. In S. Goldberg (Ed.), *Oxford Handbook of Assertion*. Oxford University Press. https://doi.org/10.1093/oxfordhb/9780190675233.013.30

Guala, F., & Hindriks, F. (2015). A Unified Social Ontology. *Philosophical Quarterly, 65*, 177–201.

Guala, F. (2016). *Understanding Institutions*. Princeton U.P.

Hart, H. L. A. (1961). *The Concept of Law*. Clarendon.

Hédoin, C. (2015). Accounting for Constitutive Rules in Game Theory. *Journal of Economic Methodology, 22*, 439–461.

Hinchman, E. (2019). Assertion and Testimony. In S. Goldberg (Ed.), *Oxford Handbook of Assertion*. Oxford University Press. https://doi.org/10.1093/oxfordhb/9780190675233.013.23

Hindriks, F. (2009). Constitutive Rules, Language, and Ontology. *Erkenntnis, 71*, 253–275.

Iseminger, G. (2006). The Aesthetic State of Mind. In M. Kieran & D. M. I. Lopes (Eds.), *Knowing Art: Essays in Aesthetics and Epistemology* (pp. 98–110). Springer.

Jacobson, D. (2011). Fitting Attitude Theories of Value. In, E. N. Zalta (ed.), *The Stanford Encyclopedia of Philosophy* (Spring 2011 Edition). https://plato.stanford.edu/archives/spr2011/entries/fitting-attitude-theories/

Jankovic, M. (2018). Collective Intentionality and Language. In M. Jankovic & K. Ludwig (Eds.), *The Routledge Handbook of Collective Intentionality* (pp. 363–377). Routledge.

Johnson, C. R. (2018). What Norm of Assertion? *Acta Analytica, 33*, 51–67.

Juul, J. (2005). *Half-Real: Video Games between Real Rules and Fictional Worlds*. MIT Press.

Kaluziński, B. (2018). What Does it Mean that Constitutive Rules Are in Force? *Argumenta, 4*(1), 111–124.

Kaplan, J. (2021). Attitude and Social Rules, or Why It's Okay to Slurp Your Soup. *Philosophers' Imprint*.

Kaufmann, M. (2012). *Interpreting Imperatives*. Springer.

Kemp, G. (1999). The Aesthetic Attitude. *British Journal of Aesthetics, 39*, 392–399.

Kennedy, C. (2007). Vagueness and Grammar: The Semantics of Relative and Absolute Gradable Predicates. *Linguistic and Philosophy, 30*(1), 1–45.

Khalidi, M. A. (2021). Are Sexes Natural Kinds? In S. Dasgupta, R. Dotan, & B. Weslake (Eds.), *Current Controversies in Philosophy of Science* (pp. 163–176). Routledge.

King, A. (2012). The Aesthetic Attitude. *Internet Encyclopedia of Philosophy.*

Kivy, P. (2011). *Once-told Tales*. Blackwell.

König, E., & Siemund, P. (2007). Speech Act Distinctions in Grammar. In T. Shopen (Ed.), *Language Typology and Semantic Description* (pp. 276–324). Cambridge University Press.

Kreider, A. J. (2011). Game Playing Without Rule-Following. *Journal of the Philosophy of Sport, 38*, 55–73.

Kretchmar, S. (2001). A Functionalist Analysis of Game Acts: Revisiting Searle. *Journal of the Philosophy of Sport, 28*, 160–172.

Kretchmar, S. (2015). Formalism and Sport. In M. McNamee & W. Morgan (Eds.), *The Routledge Handbook of the Philosophy of Sports* (pp. 11–21). Routledge.

Lehman, C. K. (1981). Can Cheaters Play the Game? *Journal of the Philosophy of Sport, 8*, 41–46.

Levinson, J. (1998). Two Notions of Interpretation. In A. Haapala & O. Naukkarinen (eds.), *Interpretation and its Boundaries* (pp. 2–21). Helsinki: Helsinki University Press. Also in his *Contemplating Art*, Oxford: OUP 2006, 275–287.

Levinson, J. (2016). Toward an Adequate Conception of Aesthetic Experience. In *Aesthetic Pursuits* (pp. 28–46). Oxford University Press.

Lewis, D. (1979). Scorekeeping in a Language Game. *Journal of Philosophical Logic, 8*, 339–359; in Lewis, D. (1983). *Philosophical Papers* (Vol. 1, pp. 233–249), Oxford: Oxford University Press.

Lewis, D. (1969). *Convention: A Philosophical Study*. Harvard U.P.

Lewis, D. (1983). New Work for a Theory of Universals. *Australasian Journal of Philosophy, 61*, 343–377.

Lind, R. (1980). Attention and the Aesthetic Object. *Journal of Aesthetics and Art Criticism, 39*(2), 131–142.

Lind, R. (1992). The Aesthetic Essence of Art. *Journal of Aesthetics and Art Criticism, 50*(2), 117–129.

Livingston, P. (2005). *Art and Intention*. Oxford University Press.

Longworth, F., & Scarantino, A. (2010). The Disjunctive Theory of Art: The Cluster Account Reformulated. *British Journal of Aesthetics, 50*(2), 151–167.

Maitra, I. (2011). Assertion, Norms, and Games. In J. Brown & H. Cappelen (Eds.), *Assertion: New Philosophical Essays* (pp. 277–295). Oxford University Press.

Marques, T., & García-Carpintero, M. (2014). Disagreement about taste: commonality presuppositions and coordination. *Australasian Journal of Philosophy, 72*(4), 701–723.

Martínez, M. (2020). Synergic Kinds. *Synthese, 197*, 1931–1946.

Matthen, M. (2017). The Pleasure of Art. *Australasian Philosophical Review, 1*(1), 6–28.

McPherson, T. (2011). Against Quietist Normative Realism. *Philosophical Studies, 154*(2), 223–240.

Midgley, G. C. J. (1959). Linguistic Rules. *Proceedings of the Aristotelian Society, 59*, 271–290.

Miller, R. (1998). Three Versions of Objectivity: Aesthetic, Moral, and Scientific. In J. Levinson (Ed.), *Aesthetics and Ethics: Essays at the Intersection* (pp. 26–59). Cambridge U.P.

Morgan, W. J. (1987). The Logical Incompatibility Thesis and Rules: A Reconsideration of Formalism as an Account of Games. *Journal of the Philosophy of Sport, 14*, 1–20.

Morgan, W. J. (2004). Moral Antirealism, Internalism, and Sport. *Journal of the Philosophy of Sport, 31*, 161–183.

Morin, O. (2011). Three Ways of Misunderstanding the Power of Rules. In M. Schmitz, B. Kobow, & H.-B. Schmidt (Eds.), *The Background of Social Reality* (pp. 185–201). Springer.

Newman, G., & Knobe, J. (2019). The Essence of Essentialism. *Mind and Language*. https://doi.org/10.1111/mila.12226

Nguyen, C. T. (2017). Philosophy of Games. *Philosophy Compass*. https://doi.org/10.1111/phc3.12426

Preston, B. (2020). Artifact. In E. N. Zalta (ed.), *The Stanford Encyclopedia of Philosophy* (Fall 2020 Edition). https://plato.stanford.edu/archives/fall2020/entries/artifact/

Putnam, H. (1975). The Meaning of 'Meaning'. In *Philosophical Papers vol. 2* (pp. 215–271). Cambridge U.P..

Ransdell, J. (1971). Constitutive Rules and Speech-Act Analysis. *Journal of Philosophy, lxviii*, 385–400.

Rawls, J. (1955). Two Concepts of Rules. *The Philosophical Review, 64*, 3–32.

Reiland, I. (2020). Constitutive Rules: Games, Language, and Assertion. *Philosophy and Phenomenological Research, 100*(1), 136–159.

Rescorla, M. (2009). Assertion and its Constitutive Norms. *Philosophy and Phenomenological Research, 79*(1), 98–130.

Ridge, M. (2017). Play and Games: An Opinionated Introduction. *Philosophy Compass*. https://doi.org/10.1111/phc3.12573

Ridge, M. (2018). The Compatibility of Games and Artworks. *Journal of the Philosophy of Games, 1*(1). https://doi.org/10.5617/jpg.5915

Ridge, M. (2020). Individuating Games. *Synthese*, online first. https://doi.org/10.1007//s11229-020-02603-5

Ridge, M. (2019). How to Play Well with Others. In T. Hurka (Ed.), *Games, Sports, and Play: Philosophical Essays* (pp. 74–97). Oxford University Press.

Roberts, C. (2018). Speech Acts in Discourse Contexts. In D. Fogal, D. Harris, & M. Moss (Eds.), *New Work in Speech Acts* (pp. 317–359). OUP.

Rough, B. (2018a). The Incompatibility of Games and Artworks. *Journal of the Philosophy of Games, 1*(1), 1–21.

Rough, B. (2018b). Response to Michael Ridge. *Journal of the Philosophy of Games, 1*(1), 1–3.

Russell, J. S. (1999). Are Rules All an Umpire Has to Work With? *Journal of the Philosophy of Sport, 26*, 27–49.

Russell, J. S. (2004). Moral Realism in Sport. *Journal of the Philosophy of Sport, 31*(1), 142–160.

Scruton, R. (1974). *Art and Imagination*. Routledge.

Searle, J. (1969). *Speech Acts: An Essay in the Philosophy of Language*. Cambridge U.P.

Searle, J. (1995). *The Construction of Social Reality*. Free Press.

Searle, J. R. (1974–1975). The Logical Status of Fictional Discourse. *New Literary History, 6*, 319–32; also in his *Expression and Meaning*, Cambridge: CUP, 1979, 58–75.

Searle, J. R. (2018). Constitutive Rules. *Argumenta, 4*(1), 51–54.

Simon, R. L. (2000). Internalism and Internal Values in Sport. *Journal of the Philosophy of Sport, 27*, 1–16.

Southwood, N. (2019). Laws as Conventional Norms. In D. Plunkett, S. Shapiro, & K. Toh (Eds.), *Dimensions of Normativity* (pp. 23–44). Oxford University Press.

Stecker, R. (2010). *Aesthetics and the Philosophy of Art* (2nd ed.). Rowman & Littlefield.

Stock, K. (2021). *Material Girl: Why Reality Matters for Feminism*. Fleet, Little Brown.

Suits, B. (1978). *The Grasshopper: Games, life and Utopia*. Scottish Academic Press; reissued in 2005, Canada: Broadview Press.

Suits, B. (1967). What is a Game? *Philosophy of Science, 34*(2), 148–156.

Thomasson, A. L. (1999). *Fiction and Metaphysics*. Cambridge U.P.

Thomasson, A. L. (2014). Public Artifacts, Intentions and Norms. In M. Franssen, P. Kroes, T. Reydon, P. Vermaas, et al. (Eds.), *Artefact Kinds: Ontology and the Human-made World*. Springer.

Tobia, K., Newman, G., & Knobe, J. (2020). Water Is and Is Not H_2O. *Mind and Language* 35, 183-208

Tyler, A., & Evans, V. (2003). *The Semantics of English Prepositions*. Cambridge University Press.

Unnsteinsson, E. (2017). A Gricean Theory of Malaprops. *Mind and Language, 32*(4), 446–462.

Walton, K. (1990). *Mimesis and Make-Believe*. Harvard U.P.

Walton, K. (1993). How Marvelous! Toward a Theory of Aesthetic Value. *The Journal of Aesthetics and Art Criticism, 51*(3), 499–510.

Walton, K. (2015a). 'It's Only a Game!': Sports as Fiction. In *In Other Shoes: Music, Metaphor, Empathy, Existence* (pp. 77–83). Oxford University Press.

Walton, K. (2015b). Fictionality and Imagination—Mind the Gap. In *In Other Shoes: Music, Metaphor, Empathy, Existence* (pp. 17–35). Oxford University Press.

Williamson, T. (1996/2000). Knowing and Asserting. *Philosophical Review, 105*, 489–523; chapter 11 of his *Knowledge and Its Limits*, New York: Oxford U.P., 2000.

Wilson, G. (1986). *Narration in Light*. The John Hopkins University Press.

Woodward, R. (2014). Walton on Fictionality. *Philosophy Compass, 9*(12), 825–836.

Index[1]

[1] Note: Page numbers followed by 'n' refer to notes.

E. Terrone, V. Tripodi (eds.), *Being and Value in Technology*,
https://doi.org/10.1007/978-3-030-88793-3

The manufacturer's authorised representative in the EU is Springer Nature Customer Service Centre GmbH, Europaplatz 3, 69115 Heidelberg, Germany. If you have any concerns regarding our products, please contact ProductSafety@springernature.com

Printed and bound by CPI Group (UK) Ltd, Croydon, CR0 4YY
22/09/2026
02221066-0003